PUBLIC HEALTH IN THE 21ST CENTURY

BIOLOGICAL CLOCKS: EFFECTS ON BEHAVIOR, HEALTH AND OUTLOOK

PUBLIC HEALTH IN THE 21ST CENTURY

Health-Related Quality of Life
Erik C. Hoffmann (Editor)
2009. ISBN: 978-1-60741-723-1

COPD Is/Is Not a Systemic Disease?
Claudio F. Donner (Editor)
2009. ISBN: 978-1-60876-051-0

Cross Infections: Types, Causes and Prevention
Jin Dong and Xun Liang (Editors)
2009. ISBN: 978-1-60741-467-4

Swine Flu and Pig Borne Diseases
Viroj Wiwanitkit
2009. ISBN: 978-1-60876-291-0

Family History of Osteoporosis
Afrooz Afghani (Editor)
2009. ISBN: 978-1-60876-190-6

Biological Clocks: Effects on Behavior, Health and Outlook
Oktav Salvenmoser and Brigitta Meklau (Editors)
2010. ISBN: 978-1-60741-251-9

PUBLIC HEALTH IN THE 21ST CENTURY

BIOLOGICAL CLOCKS: EFFECTS ON BEHAVIOR, HEALTH AND OUTLOOK

OKTAV SALVENMOSER
AND
BRIGITTA MEKLAU
EDITORS

Nova Science Publishers, Inc.
New York

For permission to use material from this book please contact us:
Telephone 631-231-7269; Fax 631-231-8175
Web Site: http://www.novapublishers.com

LIBRARY OF CONGRESS CATALOGING-IN-PUBLICATION DATA

Biological clocks : effects on behavior, health, and outlook / editors, Oktav Salvenmoser and Brigitta Meklau.

p. ; cm.

Includes bibliographical references and index.

ISBN 978-1-60741-251-9 (hardcover)

1. Biological rhythms. I. Salvenmoser, Oktav. II. Meklau, Brigitta.

[DNLM: 1. Biological Clocks--physiology. 2. Child. 3. Chronotherapy. 4. Circadian Rhythm--immunology. 5. Infant. 6. Photoperiod. 7. Time Factors. QT 167 B6127 2009]

QP84.6.B558 2009

612'.022--dc22

2009036331

Published by Nova Science Publishers, Inc. ✢ New York

CONTENTS

PREFACE

The emerging discipline denominated "chronoecology of behavior" is the result of the interaction between the work of behavioral ecologists and chronobiologists - those who explain biological rhythms on the basis of experimental studies carried out in the laboratory. Since the beginning of this century, this interaction has generated interest in understanding how biological rhythms behave in organisms living in their natural habitat. This approach requires field studies of rest/activity rhythms in order to obtain detailed hour/hour and day/day information about the possible effects of the modulation of certain environmental variables, such as photoperiod, temperature and the lunar sky, which may synchronize and/or mask behavioral activity. This book looks at the behavioral variations in non-human primates and Homo sapiens to find out how the circadian rhythm modulates behavior and in what way social interaction influences the rhythms of circadian activity. This book also analyzes the interaction between the circadian and immune systems, and explores the signal transduction pathways which could participate in this dialogue. The precise knowledge of this interaction might be extremely useful for the understanding of diseases development such as cancer. Other chapters in this book assess the quality of sleep in children and how it affects the rhythmical functions of the body. The comprehensive coverage of this topic is presented together with a general perspective from a child, caregiver and healthcare provider.

Chapter 1 - The suprachiasmatic nucleus (SCN) and the pineal gland are critical components of a neural oscillator system implicated in the timing of a wide variety of biological processes. The circadian cycles established by this biological clock occur throughout nature and have a period of approximately 24 hours. With advancing age these daily fluctuations deteriorate, leading to disrupted cycles with a reduced amplitude. In humans, age-related changes have been described for hormonal rhythms, body core temperature, sleep-wakefulness, and several other behavioral cycles. It appears that the disruption of circadian rhythms and the increased incidence of disturbed sleep during aging are paralleled by age-related alterations in the neural and temporal organization of the circadian oscillator and a decreased photic input to the clock. The observed neuronal degeneration of the SCN-pineal complex in senescence strongly suggests an organic deterioration of the circadian oscillator during aging and supports the idea that damage to the biological clock is the underlying anatomical substrate for the clinically often-observed disturbances in circadian rhythmicity in mood disorders and Alzheimer's disease. Recent studies furthermore indicate that illuminance and photoperiod may be considered as potential environmental factors controlling the functional activity of the circadian timing system. If light is the effective Zeitgeber in man, as it is in many other organisms, disturbances in processing of light information during aging or in some neurodegenerative and affective diseases may have profound effects on the timing of a variety of physiological and behavioral activities, including sleep. In view of the assumed sensitivity of the biological clock to changes in day length and light intensity research is now focussing on whether the synchronizing and antidepressant effects of bright light exposure in sleep disturbances and mood disorders can be improved by applying the therapy in temporal coherence with the circadian cycles underlying the neuronal activity of the biological clock.

Chapter 2 - To interpret seasonal timing in arthropods the change of photosensitivity to light/dark cycles during certain ontogenetic phases has been more or less ignored when modelling photoperiodic responses. The consequences of changing photosensitivity are exemplified using the large white butterfly, *Pieris brassicae*, and its response to daylength. It is an essential character in two ways: 1) Because of the 'required day number' (RDN) necessary to induce a specific response, which changes with photosensitivity. 2) Because of perception of increasing and decreasing daylength, which is not a specific ability of a species but a consequence of changing photosensitivity.

With photosensitivity-tests, evidence for a general concept of quantitative time measurement in arthropods is presented. In this context the findings reveal a converse effect of the same photoperiod depending on the light/dark regime in which an individual grows up. This opposite effect supports the argument for the existence of two independent targets for light/dark-cycles, interpreted as two antagonistic time measurement systems, and gives evidence for a 'double circadian oscillator clock' mechanism which is based on two submechanisms, a 'short-night determining system' and a separate 'long-night determining system'. The existence and independence of two systems is shown by differences in long-night vs. short-night responses regarding photosensitivity, temperature dependence, duration of diapause, and heritable factors.

Chapter 3 - There is an increasing interest to improve the wellbeing of life especially amongst the middle to upper social class in the well-developed nations. Particularly, the study on quality sleep each night has refueled much attention to the overall wellness of these individuals. Studies have shown that disturbances during sleep not only can lead to mild irritations but also adverse events and life satisfaction as a whole. At the heart of the issue, the rhythmical functions of the body can be distorted when the occurrences of sleep disturbances become frequent. Sleep is now no longer viewed simply as a state of rest but it has specific and affirmative effects. Essentially, the brain is the prime beneficiary of sleep in order to maintain its cerebral capacities. Hormonal changes under the control of the sympathetic nervous system can increase blood flow to the muscles. This promotes the breakdown of stored nutrients into glucose in muscle tissue, thus providing energy available for the body to use. Monitoring the quality of sleep has stretched beyond just for individuals with susceptible sleep disorders but for many who want to maximize their daily lives. Compound with the fact that in many well-developed countries, birth rate has been low; caregivers are becoming more concerned in bestowing prime care for the long-term development and wellness of their children. However, limited research has been initiated to assess the quality of sleep in children until the recent decades. This can be due to the misconception or poor understanding of caregivers in pediatric sleep disorders. With much recent public knowledge in this domain, there is a paradigm shift in the mindset of caregivers. Thus, the once much-neglected yet specialized field of study has evolved since, from complex studies only conducted in fully equipped sleep laboratories to possibly simplified, miniature and portable ambulatory based studies. In this chapter, the comprehensive coverage of this topic will be presented together with a general perspective from a child, caregiver and healthcare provider.

Chapter 4 - Knowledge is recently building regarding the factors that affect the development of an optimal sleep-wake cycle. Fetuses near term experience arousal fluctuations and sleep episodes already in the darkness of the womb. Yet, the factors that allow for a smooth emergence of an optimal sleep wake cycle are the conditions that place newborn at higher risk to develop sleep disorders are not yet fully understood. The current chapter proposes an integrative model for the development of circadian sleep-wake rhythms in preterm infants. This model is based on a comprehensive review of the basic science literature as well as clinical work with at-risk preterm infants. This integrative model proposed a three source risk source: 1) infant dependent neurobiological vulnerability risk, such as brainstem mediated functional disability; 2) a familial source that entails genetic predispositions, parental style and support resources; and 3) an environmental stimulation source, such as excessive NICU related stimulation and handling. These three realms act as main effects and interact with the infants' prenatal and postnatal age and with the infants' self- regulation mechanisms. The manner in which these sources operate and interact lead to directions of interventions that ameliorate the development of circadian sleep-wake rhythm deficit in infants born at-risk for sleep disorders. The chapter details each component of the model and illuminates the complexity of the interactions involved in it to deepen the understanding the mechanisms involved in early organization of sleep-wake rhythms in infants.

Chapter 5 - The daily environmental changes have imposed a selective pressure for life on Earth, driving the development of a circadian clock mechanism for the generation and entrainment of rhythms in physiological and behavioral variables (e.g. body temperature, hormonal secretion, sleep, locomotor activity, etc.). In mammals, the clock resides in the hypothalamic suprachiasmatic nuclei (SCN), and the principal signal that adjusts its activity is the light-dark cycle. Most immune factors and processes are under circadian control, although the efferent pathways that control these cycles are not completely understood. In addition, circadian disorders are usually associated with disease, and temporal disarrangements in immune parameters are closely related to the onset or the development of pathologies. In particular, several lines of evidence link the circadian clock to cancer disease. Epidemiologic studies show associations between circadian rhythms disruption and cancer risk or inadequate responses to treatment. Moreover, proinflammatory cytokines, some of which have been reported to be key actors in tumour signaling, can generate deep behavioral alterations comprising clock-associated variables during tumor progression, such as altered sleep pattern, fatigue, and lower quality of life.

The authors argue that the circadian-immune interaction operates in two directions: while the SCN (or other, peripheral, clocks) might drive circadian variations in several immune variables, humoral signals can in turn affect the molecular mechanism of the circadian clock and its entrainment. For example, higher diurnal lethality depicts a strong circadian modulation of lipopolysaccharide (LPS)-induced sepsis, which is also observed using TNF-α stimulation. Current clinical data show a strong correlation between time of day and illness manifestation or immune activity. In the other direction, administration of subpyrogenic doses of LPS, as well as of proinflammatory cytokines IL-1β and TNF-α, induces changes in SCN-driven locomotor activity rhythms. At the molecular level, this effect is achieved through activation of a transcriptional pathway involving nuclear factor-κB (NF-κB). This and other pathways (such as mitogen-activated-kinase (MAPK) signaling), lead to the expression of specific genes of the circadian clockwork, which in turns regulates its rhythmic output. Regarding cancer disease, the clock genes *Per1* and *Per2* are involved in DNA damage response, and could therefore regulate tumour suppression. Moreover, the expression of many cell-cycle genes is under circadian control. Alterations in other components of the molecular clock, as those in members of the casein kinase (CK) family, have also been linked to cancer disease.

In this review the authors analyze the interaction between the circadian and immune systems, and explore the signal transduction pathways which could participate in this dialogue. The precise knowledge of this interaction might be extremely useful for the understanding of diseases development such as cancer.

Chapter 6 - The emerging discipline denominated "chronoecology of behavior" is the result of the interaction between the work of behavioral ecologists –those who study behavioral phenomena in the natural environment– and chronobiologists, who explain biological rhythms on the basis of experimental studies carried out in the laboratory. Since the beginning of this century, this interaction has generated interest in understanding how biological rhythms behave in organisms living in their natural habitat. This approach requires field studies of rest/activity rhythms in order to obtain detailed hour/hour and day/day information about the possible effects of the modulation of certain environmental variables, such as photoperiod, temperature and the lunar sky, which may synchronize and/or mask behavioral activity. As a model for the study, a neotropical monkey species was chosen, the spider monkey, which is characterized by being widespread in America, from Mexico to southern Bolivia, and by its fusion-fission society and vast behavioral and vocal repertoire. In carrying out the studies, motor activity was registered using *Actiwatch* accelerometers during 24 continuous hours for 12 months, in combination with focal recordings of behavior and geophysical variables. On the basis of the results, the role of geophysical (astronomical and meteorological) variations as *zeitgeber* and/or the masking of the circadian rhythm, contributed valuable information for understanding alterations in physiology and behavior introduced by temporal adaptations to environmental changes. Finally, it is well-known that behavioral variations in non-human primates and *Homo sapiens* are influenced by astronomical and meteorological factors and by social, vocal, verbal and non-verbal conspecific interactions; therefore, it is essential to dissect these variations in order to study them and find out how the circadian rhythm modulates behavior and in what way social interaction influences the rhythms of circadian activity.

Chapter 7 - Self-organizing of open dynamic systems far from thermodynamic equilibrium gives rise to complexity in space and time, thus creating new structures including biological clocks. In fact, periodic oscillations can be used to control deterministic chaos. The human organism is a multioscillator, which is governed by the master clock in the hypothalamus. The authors have previously observed a periodic oscillation of free radical production that functions synchronous to the central clock and is apparently generated by it. The authors found as well, that this oscillation, possibly mediated by mitochondrial activity, ceases to function in humans suffering from advanced cancer diseases. Alterations of mitochondrial bioenergetics have been observed to be a general feature of malignant cells and could progressively diminish the distance from thermodynamic equilibrium necessary to maintain multicellular complexity. Self-organization could thus break down, leading to malignant development as a survival strategy of tumor cells.

Chapter 8 - In this chapter a “klepsydraic” model of internal time representation is introduced. The model’s working is illustrated by experimental data on duration reproduction and duration discrimination. In addition, special properties of the reproduction function predicted by the model are discussed. It is shown that time-scales generated by “klepsydraic clocks,” although non-uniform with respect to the objective (physical) time, allow for rational time-keeping. These findings recommend the model as an analytic instrument in studies of time perception, timing behavior and subjective time awareness.

In: Biological Clocks: Effects on Behavior... ISBN: 978-1-60741-251-9
Editors: O. Salvenmoser et al. pp. 1-37

Chapter 1

THE HUMAN BIOLOGICAL CLOCK: FROM GENES TO CHRONOTHERAPY

*Michel A. Hofman**
Netherlands Institute for Neuroscience Amsterdam, the Netherlands

ABSTRACT

The suprachiasmatic nucleus (SCN) and the pineal gland are critical components of a neural oscillator system implicated in the timing of a wide variety of biological processes. The circadian cycles established by this biological clock occur throughout nature and have a period of approximately 24 hours. With advancing age these daily fluctuations deteriorate, leading to disrupted cycles with a reduced amplitude. In humans, age-related changes have been described for hormonal rhythms, body core temperature, sleep-wakefulness, and several other behavioral cycles. It appears that the disruption of circadian rhythms and the increased incidence of disturbed sleep during aging are paralleled by age-related alterations in the neural and temporal organization of the circadian oscillator and a decreased photic input to the clock. The observed neuronal degeneration of the SCN-pineal complex in senescence strongly suggests an organic deterioration of the circadian oscillator during aging and supports the idea that damage to the biological clock is the underlying anatomical substrate for the clinically often-observed disturbances in circadian rhythmicity in mood disorders and Alzheimer's

* Correspondence: M.A. Hofman. Ph.D. Netherlands Institute for Neurosciene, Meibergdreef 47, 1105 BA Amsterdam. The Netherlands. Telephone number: +31-20-5665500. Fax number: +31-20-6961006. E-mail: M.Hofman@nin.knaw.nl

disease. Recent studies furthermore indicate that illuminance and photoperiod may be considered as potential environmental factors controlling the functional activity of the circadian timing system. If light is the effective Zeitgeber in man, as it is in many other organisms, disturbances in processing of light information during aging or in some neurodegenerative and affective diseases may have profound effects on the timing of a variety of physiological and behavioral activities, including sleep. In view of the assumed sensitivity of the biological clock to changes in day length and light intensity research is now focussing on whether the synchronizing and antidepressant effects of bright light exposure in sleep disturbances and mood disorders can be improved by applying the therapy in temporal coherence with the circadian cycles underlying the neuronal activity of the biological clock.

Keywords: Suprachiasmatic nucleus - Pineal gland - Human brain - Hypothalamus - Circadian oscillator - Clock genes - Biological clock - Neuroplasticity - Sexual differentiation - Aging - Alzheimer's disease.

1. INTRODUCTION

Many organisms experience circadian oscillations in various biological processes (e.g., neuroendocrine, autonomic, cardiovascular, sleep-wake cycle). Circadian rhythms enable organisms to anticipate periodic changes in the environment and are consequently important adaptive mechanisms. In mammals, these circadian cycles are regulated by an endogenous clock, the central component of which resides in the suprachiasmatic nucleus (SCN) of the hypothalamus.

Since the discovery of the SCN as the site of the master circadian pacemaker, many attempts have been undertaken to unravel the mechanisms underlying its endogenous circadian rhythmicity. In particular, lesion and transplantation experiments of the SCN and *in vitro* slice studies have provided firm evidence for its biological clock characteristics (for reviews, see Buijs et al., 1996; Van Esseveldt et al., 2000; Panda et al., 2002). Lesioning the SCN results in a disappearance of most circadian rhythms and makes the animal arrhythmic, while transplantation of fetal SCN tissue may restore circadian rhythmicity in such lesioned animals. When the SCN is removed from the brain and maintained in a slice preparation, the neurons continue to generate circadian rhythms in electrical activity, secretion and gene expression. Consistent with its role in the temporal organization of circadian processes, investigations in rodents and non-human primates suggest that the SCN is also involved in the seasonal timing of reproduction, sexual behavior and energy metabolism (for a review, see Hofman, 2004).

Studies in humans also seem to support the notion of the SCN being the principal neural substrate that organizes and coordinates circadian rhythms. Clinically documented disruption of circadian behavior shows involvement of the SCN region (Schwartz et al., 1986; Cohen and Albers, 1991) and age-related decrements in circadian timekeeping have been attributed to the observed neuronal degeneration of the SCN in senescence (Swaab et al., 1985, 1996; Moore, 1991; Hofman, 2000; Hofman and Swaab, 2006). It appears that photic information may have a synchronizing effect on the clock mechanism of the SCN by inducing changes in the functional activity of groups of neurons.

Increasing age and a variety of diseases, however, may impair many of these functions and may have deleterious effects on the neuronal organization and biological activity of the clock. In humans, age-related changes have been described for hormonal rhythms, body core temperature, sleep-wakefulness, and several other behavioral cycles. It appears that the disruption of circadian rhythms and the increased incidence of disturbed sleep during aging are paralleled by age-related alterations in the neural and temporal organization of the circadian oscillator and a decreased photic input to the clock.

The observed neuronal degeneration of the SCN-pineal complex in senescence strongly suggests an organic deterioration of the circadian oscillator during aging and supports the idea that damage to the biological clock is the underlying anatomical substrate for the clinically often-observed disturbances in circadian rhythmicity in mood disorders and Alzheimer's disease (Hofman, 2000; Skene and Swaab, 2003; Hofman and Swaab, 2006). Recent studies furthermore indicate that illuminance and photoperiod may be considered as potential environmental factors controlling the functional activity of the circadian timing system. In this chapter, recent data on the circadian organization of the SCN-pineal complex in humans are discussed in relation to aging and Alzheimer's disease.

2. Organization of the Human SCN

2.1. Neuronal Organization

The human SCN is a small collection of parvocellular neurons in the basal part of the anterior hypothalamus, just dorsal to the optic chiasm on either side of the third ventricle (Figure 1). The bilateral SCN in humans is about 1 mm^3 in volume and contains a total number of neurons close to 100,000 (Hofman et al., 1988; Hofman and Swaab, 2002). Based on differences in morphology, afferent inputs and output projections, the SCN can be divided into a dorsomedial part or "shell" and a ventrolateral part or "core" (Mai et al., 1991; Hofman et al., 1996; Moore and Silver, 1998; Antle and Silver, 2005). Neurons in the dorsomedial region of the SCN, for example, are small and poor in dendritic arbor and contain arginine vasopressin (AVP) and neurotensin, whereas those in the ventrolateral region of the nucleus are relatively large with extensive denditric arbors and contain vasoactive intestinal polypeptide (VIP), gastrin-releasing peptide, neurotensin, neuropeptide-Y, substance-P and calbindin (Mai et al., 1991; Moore, 1992; Romijn et al., 1999; Moore et al., 2002; Figure 2). It has been established that both AVP- and VIP-expressing neurons send efferents to the rest of the brain, constituting the main output pathway of the clock (Dai et al., 1997; Buijs and Kalsbeek, 2001). Many neurons in both regions of the SCN contain γ-aminobutyric acid (GABA), an inhibitory neurotransmitter in the brain (Moore et al., 2002; Saper et al., 2005).

The ventrolateral subdivision of the SCN receives retinal input from the retino-hypothalamic tract (RHT) and secondary visual projections from the intergeniculate leaflet of the lateral geniculate complex via the geniculo-hypothalamic tract (GHT). The RHT terminals contain glutamate, as well as pituitary adenylate cyclase-activating polypeptide, which code chemically for 'light' and 'darkness' information, respectively, while the GHT has been characterized in several rodents by the neuropeptide-Y and GABA content of its neurons (Van Esseveldt et al., 2000; Shirakawa et al., 2001; Hannibal, 2002; Moore et al., 2002).

2.2. Functional Organization

The 24 h oscillations of biological processes observed in a broad spectrum of organisms are controlled by internal clocks. Although these rhythms are generated endogenously, they do not function in isolation from their surroundings. Rather, they are entrained by external cues so they can anticipate the natural temporal variations of the environment. Classically, the expression of circadian rhythms is thought to have three fundamental components: input pathways that transmit environmental cues to the circadian clock, the clock itself, which generates and coordinates the biological rhythm(s) and output pathways that transmit the clock's information regarding phase and periodicity to the rest of the brain and body (Buijs and Kalsbeek, 2001; Kalsbeek and Buijs, 2002, 2006; Hofman, 2004).

The biological clock is synchronized to the environmental light-dark cycle by photic information from the retina to the SCN. These afferent fibers predominantly terminate in the ventral part of the SCN, where they make synaptic contacts with VIP neurons (Reuss, 1996). The VIP subdivision of the SCN is therefore thought to play a prominent role in the mediation of photic information to the circadian timing system. The direct neuronal pathway from the retina to the SCN also exists in humans, as was shown by staining degenerating neurons in patients who had incurred optic nerve damage prior to death (Sadun et al., 1984) and by *in vitro* postmortem tracing studies (Dai et al., 1998).

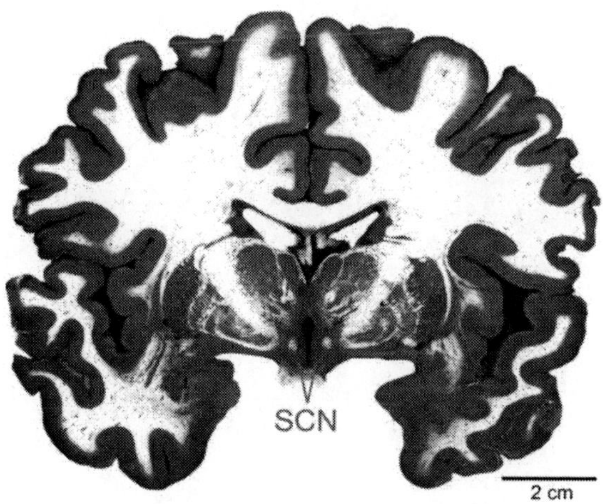

Figure 1. Coronal section through the human brain at the rostral end of the diencephalon to show the anterior part of the hypothalamus with the suprachiasmatic nucleus (SCN).

2.3. Temporal Organization

In humans and other species living under semi-natural lighting conditions, little is known about the role of VIP neurons in the temporal organization of the biological clock. Totally blind people often lack the entraining effects of light, as do many subterrestrial species, and may show free-running temperature, cortisol and melatonin rhythms. They may also suffer from sleep disturbances (Sack et al., 1992). Surprisingly, some blind people maintain circadian entrainment and show light-induced suppression of melatonin secretion, despite the apparent total lack of pupillary light reflexes and with no conscious perception of light (Czeisler, 1995). It has been proposed that in these patients the retino-hypothalamic pathway that passes through the SCN would still be intact, but that the exact nature of the circadian photoreception in these patients is unknown. It might be of practical importance to distinguish these patients, since in this group enucleation might cause recurring insomnia and other symptoms associated with the loss of circadian rhyhthmicity.

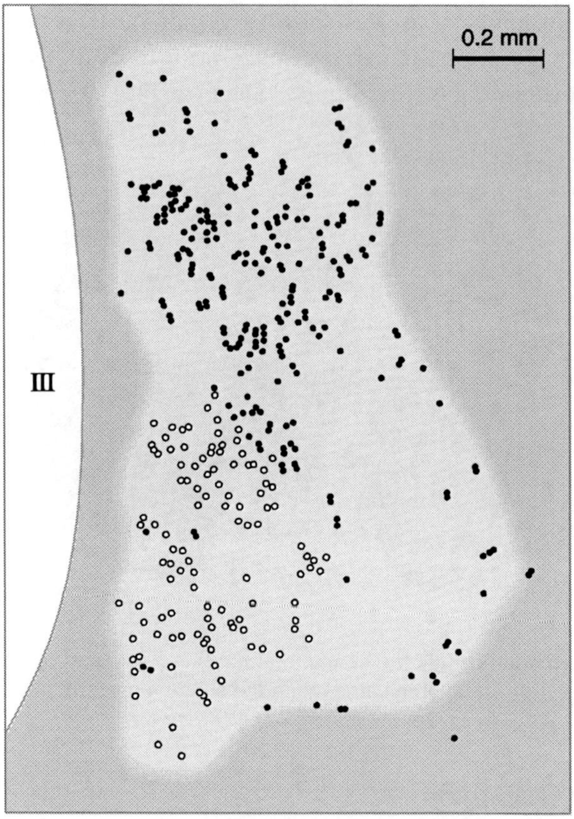

Figure 2. Distribution of arginine vasopressin (AVP)-expressing and vasoactive intestinal polypeptide (VIP)-expressing neurons in the human suprachiasmatic nucleus (SCN). Drawing showing the distribution of the AVP neurons (dots) and VIP neurons (open circles) through the mid-portion of the SCN. III = third ventricle. Modified from Hofman (2000).

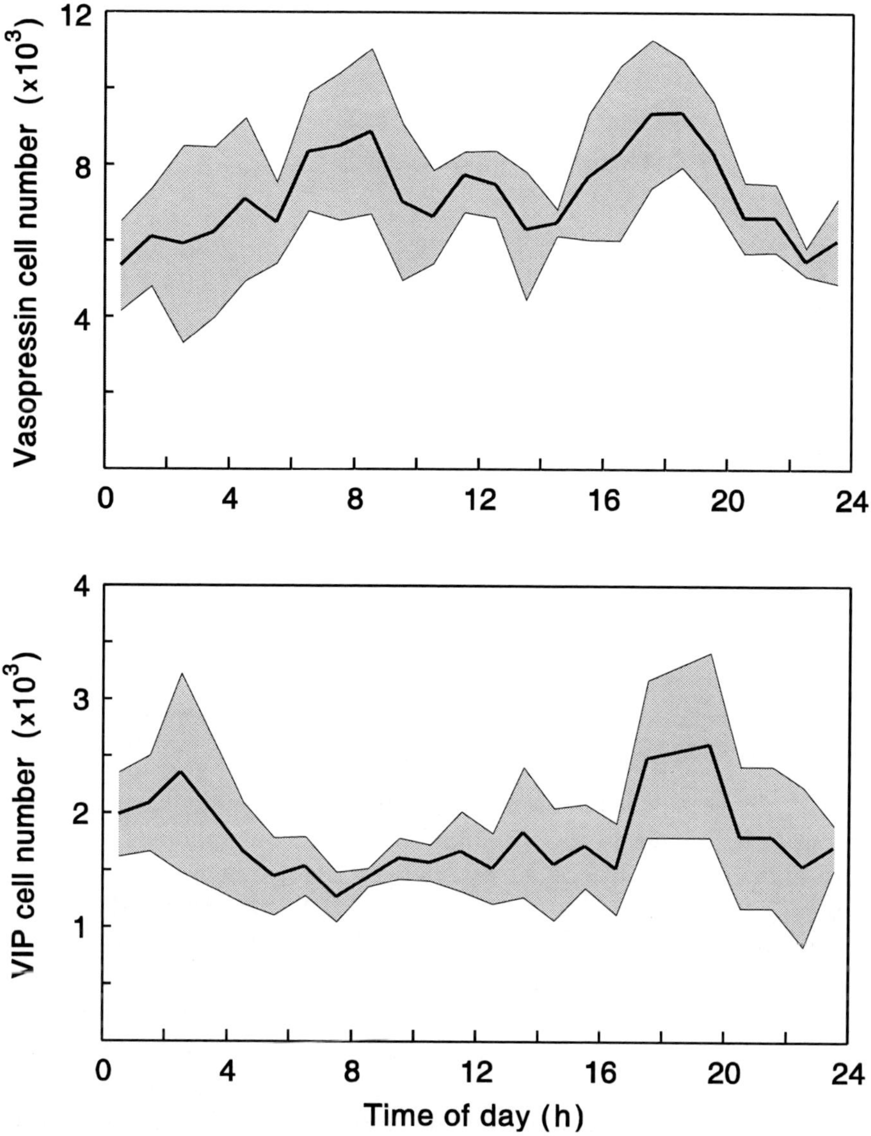

Figure 3. Circadian oscillations in the biological activity of the AVP- and VIP-expressing neurons in the human suprachiasmatic nucleus (SCN). Note the asymmetrical, bimodal waveform of the cycles. The data are represented by mean±SEM values. Reproduced from Hofman (2003).

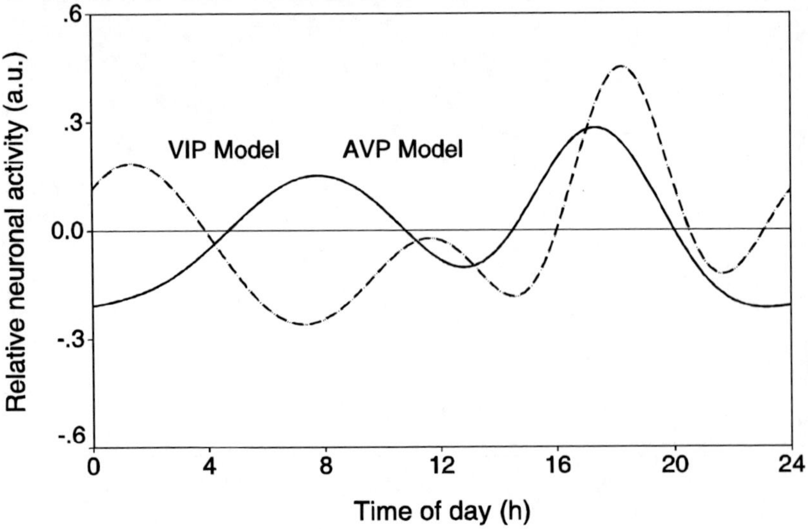

Figure 4. The nonlinear periodic functions describing the circadian AVP and VIP cycles in the human suprachiasmatic nucleus (SCN). The model of the AVP cycle reaches an estimated maximum at the end of the afternoon (tmax = 17.3 h) and a minimum around midnight (tmin = 23.2 h). The model of the VIP cycle also reaches its maximum at the end of the afternoon (tmax = 18.2 h), but has its minimum in the early morning (tmin = 7.3 h), considerably later than the minimum of the AVP cycle. Reproduced from Hofman (2003).

In the past decade, several studies have been conducted to determine whether variations in light intensity and photoperiod affect the morphology and neuronal activity of the mammalian SCN, by studying brains obtained at autopsy (for reviews, see Hofman and Swaab, 1992, 2002; Swaab et al., 1996). The AVP- and VIP-expressing neurons in the human SCN, for example, were found to exhibit distinct circadian rhythms with an asymmetrical, bimodal waveform (Hofman and Swaab, 1993; Hofman, 2000, 2003; Figure 3). The AVP cycle has a peak in the early morning, a lower plateau during the day, a second peak in the late afternoon, and a decline beginning in the early evening, leading to a nadir around midnight. The VIP cycle shows a peak in the middle of the night, a lower plateau beginning in the late night and lasting for about 12 hours, and a second peak in the late afternoon, followed by a sharp decline in the early evening.

Time series analysis, furthermore, indicated that the circadian cycles in the SCN can be adequately described by a model consisting of nonlinear periodic functions, which could be decomposed into mono- and diphasic cycles, with periods of 24-hours and 12-hours, respectively (Hofman, 2003; Figure 4). The demonstration of two significantly different, but temporally linked, output profiles strongly suggest that the circadian clock in the SCN consists of a multioscillator system.

2.4. Multioscillator Circadian System

While it is still difficult to come up with a definitive "circuit diagram" for the biological clock, a recently proposed theory, based on anatomical and biochemical studies, considers the clock as a multioscillator system, consisting of coupled, independently operating subdivisions located in the core and shell of the SCN (Moore and Silver 1998; Leak and Moore 2001). The inputs and outputs of these two regions also appear to be distinct, indicating that the nature of the signal conveyed to areas receiving core and shell projections varies as a function of the subdivision from which the innervation is derived.

It is likely that the period of the rhythmic output from the core and shell is the same, as the distinct effector sites of these subdivisions need to be entrained to the same sidereal time. By contrast, differential inputs to the core and shell should control phase, amplitude, or waveform of their rhythms separately. The specific neurological organization and information flow in the biological clock might explain why the circadian profiles of the AVP are somewhat different from those of the VIP cycle. Although the exact nature of the biological clock and the means by which it is synchronized to the 24-hour photic cycle is still elusive, the mathematical models describing the circadian rhythms in the human clock suggest that both the core and the shell of the SCN contain circadian oscillators. From rodent studies we know that the dorsomedial part of the SCN indeed contains a circadian pacemaker. The ventromedial part of the SCN, on the other hand, receiving primary and secondary visual afferents, may reflect the external light-dark cycle, rather than endogenous oscillator activity.

The notion of separate circadian pacemakers in the SCN was recently supported by a differential entrainment of two components in the circadian rhythm of SCN neuronal activity to the light-dark cycle (Jagota et al., 2000). Also the circadian rhythm of *c-fos* expression and that of *Per1* and *Per2* expression, two clock genes implicated in the molecular mechanisms of the circadian pacemaker, were found to be slightly different between the dorsomedial and ventromedial regions of the SCN (Schwartz et al., 2000; Dardente et al., 2002). It does not necessarily mean that the AVP- and VIP-expressing cells in the SCN are pacemaker cells. In fact, the role of neurotransmitters and neuromodulators in the generation of circadian rhythms is unclear. So far, none of the peptides found in the SCN has been reported to play a crucial role in the circadian oscillator system. Instead, some of the peptide genes may act as clock-controlled genes and their transcripts may drive the circadian output of the clock (for reviews, see Dunlap 1999; Cermakian and Sassone-Corsi 2000; Hastings and Maywood, 2000; King and Takahashi, 2000; Hastings et al., 2003). In other words, circadian oscillations in neuropeptide levels in the SCN are probably part of the output system of the circadian clock, and are not an intrinsic element of the circadian pacemaker.

It is evident that there must be some kind of coupling for the two components of the oscillating system in the SCN to maintain their mutual phase relationship under free-running conditions. There are various ways how the coupling of these oscillators may be envisaged, depending on whether the oscillators are restricted to different populations of neurons in the SCN or that they are colocalized in the same neuron. In the coupling interaction, it is likely that there is some degree of asymmetry in the strengths of the two oscillators (see e.g., Daan et al., 2001; Wehr, 2001; Hofman, 2003). Cell-to-cell communication is clearly necessary for conveying inputs to and outputs from the SCN and may be involved in ensuring the high precision of the observed rhythms. Although at present the details of how time of day is encoded by the molecular clockwork are not known it is clear that there are properties in the behaviors of the clock genes that could be interpreted as circadian oscillators.

From a functional perspective the main attraction of the dual oscillator model is that it uniquely counts for the adjustment of circadian organization to season and latitude and that the phase relationship of these oscillators or the temporal organization of the expression of their genes reflect the photoperiod, and thus time of year, to which an animal has been exposed (Daan et al., 2001; Hofman, 2003; 2004). This dual oscillator model also has obvious implications for the evolution of circadian timing. The rotation of the earth provides two precise timing signals, sunrise and sunset, rather than a single one. It might well be that life that evolved on its surface has developed ways to exploit both signals, for the fine tuning of temporal organization over the 24-hours, as well as for adapting its physiology and behavior to the seasonal fluctuations in photoperiod.

3. MOLECULAR GENETICS OF CIRCADIAN TIMING

3.1. Clock Genes and the Suprachiasmatic Nucleus

As internal systems, circadian clocks arise from the expression and temporally regulated activity of specific genes and gene products. The past decade has witnessed rapid breakthroughs in our understanding of how the circadian clock functions at the cellular and molecular levels (for reviews, see King and Takahashi, 2000; Reppert and Weaver, 2001; 2002; Hastings and Herzog, 2004). Individual SCN neurons express self-sustained circadian oscillations driven by autoregulatory transcription-translation feedback loops. The period of this process is approximately 24 hours, giving rise to the circadian rhythm. Beginning with the molecular cloning of the mouse *Clock* gene at least six genes now have been identified as putative components of the mammalian clock: *Per1*, *Per2*, *Cry1*, *Cry2*, *Clock* and *Bmal1*. Additional genes potentially involved in output pathways from and input pathways to the circadian clock have been identified. Oscillations of mRNA transcripts of these clock genes have been an important element of the study of biological clocks.

Based on the oscillation patterns of the *Period* genes within the SCN (high during the day, low during the night), the mammalian clock was predicted to be a "day-active" clock, in which the oscillating components of the clock are at higher levels during daytime than during the night. However, analysis of the phase response curve (PRC) to light shows that the circadian clock is more greatly affected at nighttime than during the day (daytime is within a "dead zone" during which light exposure has little or no effect on the phase of the clock) (see e.g. King and Takahashi, 2000). Thus, during the dead zone of the PRC, the oscillating clock molecules are abundant, and during the active phase of the PRC, these molecules are scarce. What follows from this is the prediction that this clock would be phase shifted (and thus entrained) not by degradation of clock molecules following zeitgeber exposure but rather by the induced expression of clock molecules.

Although phase information is essential for a functioning clock, multiple clock components may work in concert to invest this phase information in the cycling activity of one or few elements of the system. In this scenario, all of the components are important for the presence of a circadian rhythm and thus should be considered as elements of the clock, even though only a subset directly reflects the current phase of the clock. Such a scenario appears to be the case for the mammalian circadian clock, in which mRNA abundance of some components exhibit high-amplitude circadian oscillations while others have little or no apparent oscillations over the 24-hour cycle. The nearly 24-hour oscillation is thought to depend on delays between transcription, translation, and subsequent transcriptional repression, but how these delays are established is only beginning to be understood. Insights into the steps modulating synthesis, turnover, and localization of key mRNAs and proteins will inevitably modify the model for molecular timekeeping. The molecular loop must broadcast its timing signal to the SCN neuron and beyond that to brain and body. This is achieved, in part, by the temporally regulated expression of a series of clock-controlled genes.

3.2. Organizing Principles of Cellular Clocks

The human SCN is not simply a collection of 100,000 clock cells: rather, it is a heterogeneous structure composed of multiple functional compartments. Indeed recent studies challenge the view that the SCN is functionally homogeneous (see e.g., Hastings and Herzog, 2004; Antle and Silver, 2005). Some SCN cells are not endogenously rhythmic with respect to clock gene expression; only some cells receive direct retinal input and immediately express clock genes following photic stimulation. Furthermore, individual clock cells within the SCN exhibit various phases and free-running periods (Nakamura et al., 2001; Quintero et al., 2003; Yamaguchi et al., 2003). These phenomena are difficult to explain by coupling alone. An alternative 'network' model suggests that it is precisely this heterogeneity of the SCN that is integral to keeping the population of oscillators ticking in a coherent fashion (Antle et al., 2003). This model hypothesizes that the phases of individual oscillators are synchronized by a daily signal from a subset of SCN cells. By pulling the phases closer together, the overall output of the system remains rhythmic.

Central to this model are recent discoveries demonstrating that the morphological and peptidergic heterogeneity of the SCN reflects its functional heterogeneity. Rather than being uniformly distributed in the SCN, these intrinsically rhythmic cells are largely confined to the SCN shell, occupying roughly the same area as AVP-containing cells (Hamada et al., 2001; 2004; Yan and Okamura, 2002;). This region of the SCN receives little retinal innervation and displays delayed clock gene expression following phase-shifting light exposure (Hamada et al., 2001).

Cells in the SCN core receive direct retinal innervation and express c-fos, *Per1* and *Per2* in response to phase-shifting light pulses (Romijn et al., 1996; Yan et al., 1999; Hamada et al., 2001; Yan and Silver, 2002). It has been observed that cells in the core of the mouse SCN either lack rhythmic expression of *Per1* and

Per2 (Karatsoreos et al., 2004) or express these genes in antiphase to the expression in the SCN shell (King et al., 2003). Cells in the core oscillate in their responsiveness to photic input: although light exposure always increases firing rates in SCN neurons (Meijer et al., 1992), light induces clock gene expression in the SCN only during the night (Reppert and Weaver, 2001).

Although expression in each individual cell peaks at a particular time of the day that is stable for that cell over several cycles, peak expression for the population as a whole occurs during the middle of the day. When the dorsal third of the nucleus is surgically separated from the ventral portion that contains the core, coordinated rhythmic output persists in only the ventral portion (Yamaguchi et al., 2003). Individual cells in the dorsal portion remain rhythmic but coherent phase relationships among cells are lost, resulting in a tissue without net rhythmic output.

3.3. Clock Gene Expression Patterns

For the circadian clock to be phase-shifted or entrained by light, photic information must be relayed from the lightinduced cells to the rhythmic cells, which must respond appropriately to the resetting signal (with delays and advances to early-night and late-night light exposure, respectively). Detailed examination of the temporal and spatial patterns of gene expression reveal that *Per1* and *Per2* are regulated separately in a phase-specific and region-specific manner (Yan and Silver, 2002; 2004). Following phase-shifting light pulses, *Per* expression is first induced in the SCN core (Yan et al., 1999; Hamada et al., 2001; 2004; Kawamoto et al., 2003). In the mouse, early-night delaying light pulses induce *mPer1* and *mPer2* expression initially in the SCN core, followed later by *mPer2* expression in the SCN shell. Late-night advancing light pulses induce expression of *mPer1* but not *mPer2*. This *mPer1* expression occurs first in the core and later in the shell.

Mid-night light pulses that do not alter circadian phase induce *mPer1* expression only in the SCN core. Based on these findings, it appears that the spread of *mPer2* expression to the shell underlies phase delays, whereas a similar spread of *mPer1* underlies phase advances (Yan and Silver, 2002; 2004). Phase delays to light are impaired in *mPer2* mutant mice (Albrecht et al., 2001; Spoelstra et al., 2004) and following *Per2* antisense treatment (Wakamatsu et al., 2001). The role of *Per1* in light-induced phase shifts is less clear, although *Per1* antisense treatment also attenuates phase delays.

Cells in the SCN shell have a specific spatial arrangement. Daily rhythmic expression of *Per1*, *Per2* and AVP does not occur in all cells simultaneously, but rather spreads and recedes through the SCN (Hamada et al., 2004). Rhythmic expression of *Per1*, *Per2* and AVP starts in a small group of cells located adjacent to the third ventricle in the dorsomedial SCN and spreads ventrolaterally over 4–8 hours. It then recedes until expression is again limited to the dorsomedial region. The same pattern of expression has been observed in vitro using transgenic mice in which luciferase reports *Per1* expression (Yamaguchi et al., 2003). Expression initially occurs in the dorsomedial periventricular SCN and spreads slowly to the ventral SCN over several hours.

Blocking protein synthesis or electrical activity reveals functional aspects of SCN organization. Because the intracellular clock is driven by an autoregulatory transcription–translation feedback loop, individual clocks can all be reset to a common phase by a protein synthesis inhibitor. When the inhibitor is removed, cells begin to oscillate in phase with one another, but re-establish their original phase relationships after several cycles. Furthermore, maintaining this phase relationship requires electrical signaling, either between individual cells or between the core and the shell. When action potentials are blocked with tetrodotoxin, circadian patterns of luminescence in individual cells persist but the system as a whole becomes desynchronized (Yamaguchi et al., 2003). These data reveal that, although individual cells within the SCN can sustain oscillations, the specific phases of such oscillations are regulated by the network properties of the SCN.

4. The SCN-Pineal Complex

4.1. Circadian Timing

The pineal gland is a central structure in the circadian timing system and the major source of the hormone melatonin (Arendt, 1995). The pineal is innervated by a neural multi-synaptic pathway originating in the suprachiasmatic nucleus (SCN) of the hypothalamus (Buijs and Kalsbeek, 2001; Figure 5).

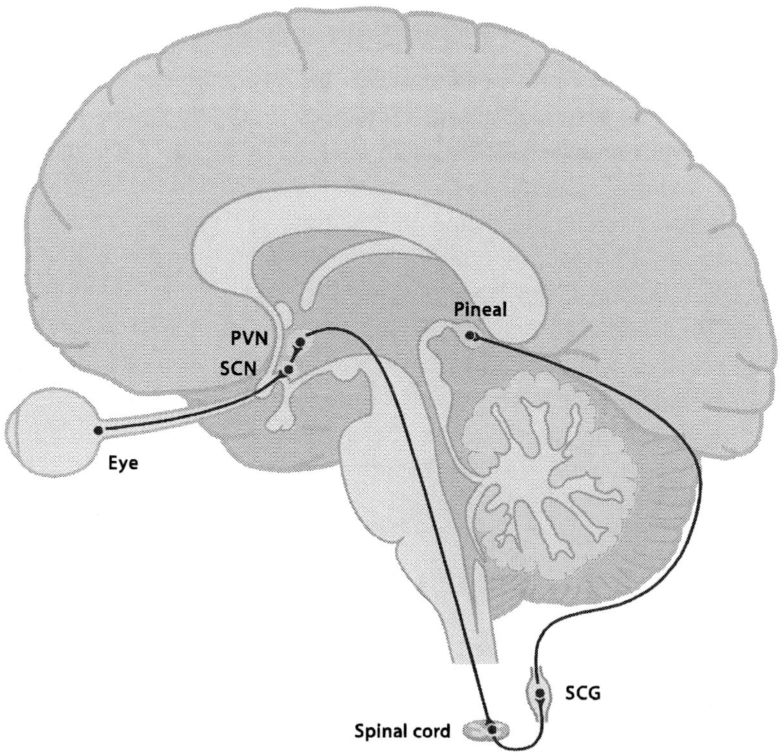

Figure 5. Diagram of the human brain (mid-sagittal section) showing the multi-synaptic pathway of the circadian timing system by which environmental light information reaches the SCN and the pineal gland. SCN = suprachiasmatic nucleus; PVN = paraventricular nucleus; SCG = superior cervical ganglion. Modified from Hofman and Swaab (1992).

Disruption of any portion of this pathway from the SCN to the pineal gland abolishes melatonin rhythmicity. In all animals studied to date, whether they are diurnal or nocturnal, there is a day/night variation in pineal melatonin production with peak concentrations occurring during the dark phase. The circadian clock and its output rhythms are synchronized to the 24 h light/dark cycle by ocular light which is transmitted from the retina primarily via the retinohypothalamic tract (RHT) to the SCN. Although functions of this hormone in humans are mainly based on correlative observations, there is some evidence that melatonin stabilises and strengthens coupling of circadian rhythms, especially of core temperature and sleep-wake rhythms (Skene and Swaab, 2003; Claustrat et al., 2005).

4.2. Seasonal Timing

In recent years it has also become clear that in mammals the SCN and the pineal gland are the principal neural structures involved in the regulation of annual cycles (for reviews see, Hofman, 2004; 2009a). In fact, many of the functions that exhibit seasonal cycles in mammals, such as sexual behavior, energy metabolism, food intake, and hibernation, are regulated by this timing system in the brain. Photoperiodic information has been shown to be the strongest synchronizer of seasonal functions in most species. These findings strongly suggest that the endocrine activity of the mammalian pineal is under neural control, and receives a major input from the SCN. This means that in addition to its role as a circadian pacemaker, the SCN may also be involved in the seasonal timing of a number of physiological and behavioral processes by regulation of the photoperiod-dependent changes in melatonin secretion.

5. Biological Clock and Aging

5.1. Circadian Rhythmicity and Aging

With advancing age the circadian timing system is progressively disturbed, both in humans and other mammals, as is clearly demonstrated by a reduced amplitude and period length of circadian rhythms and an increased tendency towards internal desynchronization (for reviews, see Van Someren, 2000a). In humans, age-related changes have been described for hormonal rhythms, body core temperature, sleep-wakefulness, and several other behavioral cycles (see, e.g., Cornélissen et al., 1992; Monk and Kupfer, 2000; Touitou and Haus, 2000; Van Someren et al., 2002; Avidan, 2005). Weitzman and colleagues (1982), for example, who studied young and elderly subjects under conditions of temporal isolation, reported a reduction in amplitude and period of body temperature rhythm.

A similar reduction in the amplitude of the circadian sleep-wake cycle was found in long-term registrations of activity patterns in young and elderly men (Renfrew et al., 1987), and even more so in patients with Alzheimer's disease (Witting et al., 1990; Van Someren et al., 1996; Bhatt et al., 2005). The increase in time spent in both wakefulness during the night and daytime naps, a characteristic pattern in many elderly people, is a symptom of disruption of circadian sleep rhythms similar to the one found following experimental SCN lesions (Eastman et al., 1984). In fact, disorders of the circadian timing system during aging may first be manifested as sleep-wake pathologies (Turek et al., 1995; Hofman, 1997; Dijk et al., 2000; Van Someren, 2000b).

The question now is why the timing and consolidation of sleep and other physiological rhythms change so greatly with age? Observations in animals suggest that there are age-related changes in the properties of the circadian oscillator, which regulates the timing of sleep, body temperature and other variables within the 24-hour day. In old animals the circadian pacemaker oscillates more rapidly and with a lower amplitude than in young ones (Morin, 1988; Satinoff et al., 1993). Moreover, a change in the molecular responsiveness of the pacemaker to photic stimulation has been found in aged rats (Sutin et al., 1993).

5.2. Clock Genes and Aging

Recent experiments in rodents show that the age-related deterioration in circadian pacemaker function occurs at the level of individual cells in the SCN, possibly accounting for age-related deficits observed in the expression of behavioral rhythmicity (Aujard et al., 2001). Indeed aging is found to be correlated with decreased expression of several of the genes that make up the molecular machinery of the circadian clock, including both *Clock* and its binding partner *Bmal1* (Asai et al., 2001; Weinert et al., 2001; Kolker et al., 2003; 2004).

Kolker and colleagues, for example, reported that age alters the 24-hour expression profile of *Clock* and *Bmal1* in the SCN of hamsters (Kolker et al., 2003). However, there was no effect of age on the 24-hour profile of either *Per1* or *Per2* in constant darkness, although light induced less *Per1,* but not *Per2* in the SCN of old hamsters. Given the importance these genes have in the generation of circadian rhythms one might expect that expression of either or both of these *Per* homologs would be decreased in the SCN of old animals. Asai et al. (2001) found similar results in rats, while Yamazaki et al. (2002) found that advanced age does not dampen rhythmic expression of a *Per1*-driven luciferase in the SCN of rats. It seems that while induction of the *Per* homologs by light are altered in old age, the daily rhythm of expression of these genes is similar in the SCN of young and old rodents.

Weinert and colleagues, who investigated the expression of clock genes in the SCN of young and old laboratory mice found that the expression of *mPer1* and *mPer2* was rhythmic in both groups, with peak values at CT17 (Weinert et al., 2001). The levels of *mClock* and *mCry1* were not different depending on the time of day and did not vary with age. In contrast, an age-dependent difference was observed in the case of *mPer2* (but not *mPer1*) expression, with the maximum at CT17 significantly lower in old mice. Clock gene expression in peripheral tissues (liver and heart) also seems to be affected by aging (Claustrat et al, 2005). The liver of old rats exhibits a significant lower level of *Per* gene expression in the evening compared to middle-aged rats. In the heart of old rats similar 24-hour gene profiles have been observed, with a tendency towards a decrease of *Per* expression and an increase of *Bmal1* expression in the evening.

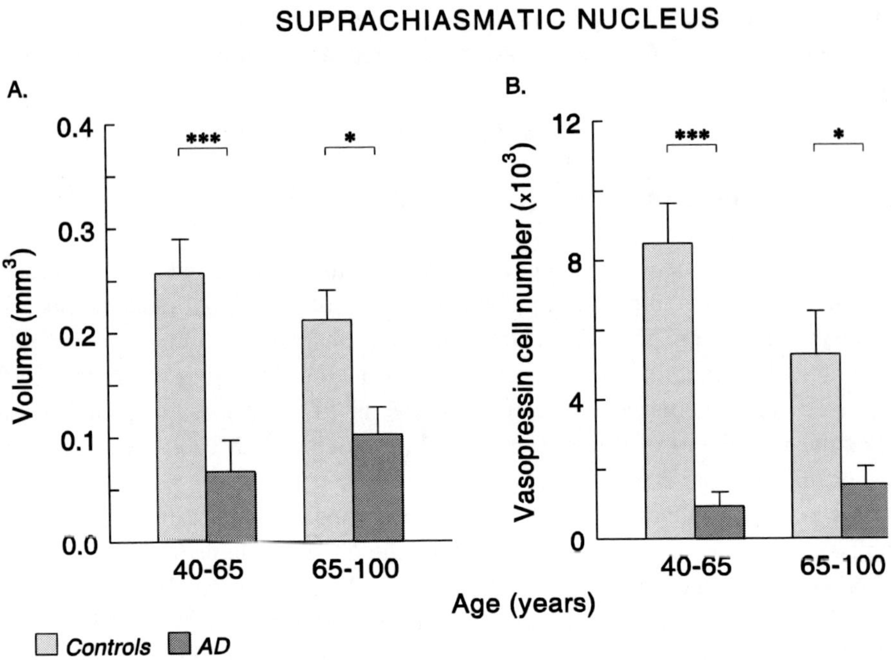

Figure 6. Effect of aging and Alzheimer's disease on the human suprachiasmatic nucleus (SCN). In Alzheimer patients, both presenile (<65 years of age) and senile (>65 years of age), the volume of the SCN (A) and the number of arginine vasopressin (AVP)-expressing neurons in the SCN (B) are significantly decreased compared to age-matched controls. Note that in presenile Alzheimer patients the number of AVP-expressing neurons is only 10% of that of controls of comparable age. Reproduced from Swaab (2004), with permission.

Furthermore, mutation experiments show that age-related changes in circadian rhythmicity occur equally in wild-type and heterogezygous *Clock* mutant mice (Kolker et al., 2004), suggesting that the *Clock* mutation does not render mice more susceptible to the effects of age on the circadian pacemaker. While they clearly have altered circadian rhythms as young adult animals, and these disruptions are exacerbated by age, the age-related changes in the rhythms of mutant mice are similar to those of wild-type mice. Together these data suggest that a mutation that has pronounced effects on the circadian timing system of young adult animals does not necessarely have a major effect on the way the circadian rhythm of locomotor activity changes with age.

5.3. Suprachiasmatic Nucleus Dysfunction in Older People

The many lines of evidence of age-related decrements in circadian time-keeping in human beings (Vitiello and Prinz., 1990; Czeisler et al., 1992; Van Someren et al., 2002) and the observed neuronal degeneration of the SCN in senescence and Alzheimer patients (Swaab et al., 1985, 1996; Zhou et al., 1995; Hofman et al., 1996; 2006; Hofman, 1997; Swaab, 1999) strongly suggest an organic deterioration of the circadian oscillator.

The disruption of circadian rhythms and the increased incidence of disturbed sleep in humans during aging (Van Someren et al., 2002) are paralleled by age-related alterations in the neural organization of the SCN, a decreased photic input to the clock, and, in the SCN of Alzheimer patients, also with a dramatic decrease in peptide synthesis (Liu et al., 2000; Swaab, 2004; Figure 6), the presence of pretangles (Swaab et al., 1992; Van de Nes et al., 1998) and tangles (Stopa et al., 1999). However, diffuse amyloid plaques are only seldom noted in this nucleus (Van de Nes et al., 1998; Stopa et al., 1999). Stopa and colleagues (1999), for example, reported that the SCN of Alzheimer patients, who are frequently suffering from sleep disruption and other circadian rhythm disturbances, is severely damaged and exhibits a dramatic loss of vasopressin- and neurotensin-expressing neurons, as well as a corresponding increase in the GFAP (glial fibrillary acidic protein) stained astrocytes.

The immunocytochemical data showing decreased activity of the SCN in Alzheimer's disease (AD) have been confirmed by *in situ* hybridization (Figure 7). The total amount of AVP mRNA in the SCN of Alzheimer patients was three times lower than in age- and sex-matched controls. In addition, the AVP mRNA-expressing neurons in the SCN showed a marked day–night difference in controls under 80 years of age. The amount of AVP mRNA was more than three times higher during the day than at night, whereas no clear diurnal rhythm of AVP mRNA was observed in AD patients (Liu et al., 2000). These data support the idea that damage to the SCN is the underlying anatomical substrate for the clinically often-observed disturbances in circadian rhythmicity in Alzheimer's disease.

Moreover, we found that the number of AVP-expressing neurons in the human SCN exhibits a marked diurnal oscillation in young (up to 50 years of age), but not in elderly people (over 50 years of age) (Hofman and Swaab, 1994; Figure 8). Whereas in young subjects low AVP-expressing neuron numbers were observed during the night and peak values during the early morning, a disrupted cycle with a reduced amplitude was found in the SCN of elderly people. Similar age-related decrements have been reported for the seasonal timing system (Hofman and Swaab, 1995).

As the number of AVP-expressing neurons probably reflects the peptidergic activity state of the cells, these findings suggest that the synthesis of AVP in the human SCN exhibits diurnal as well as seasonal fluctuations, and that the temporal organization of these rhythms becomes progressively disturbed in senescence.

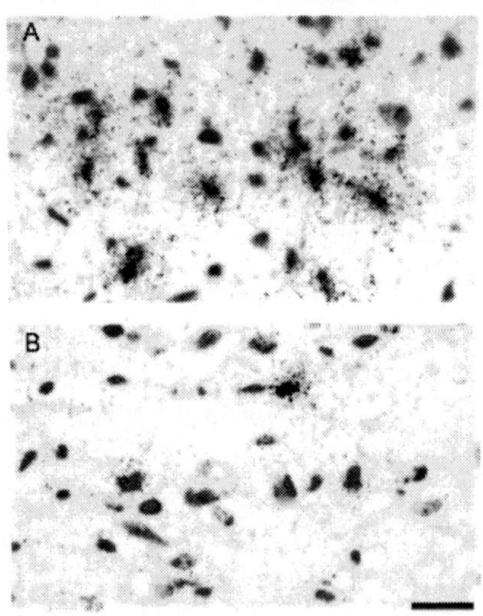

Figure 7. Thionin-counterstained emulsion autoradiograms in frontal sections (6 μm) of the human suprachiasmatic nucleus (SCN) at high magnification. Note the black silver deposits that show the presence of vasopressin (AVP) mRNA in the SCN neurons. A. Control subject. B. Alzheimer patient. Note the low number of AVP mRNA expressing neurons in the AD patient. Scale bar = 100 μm. Modified from Liu et al. (2000).

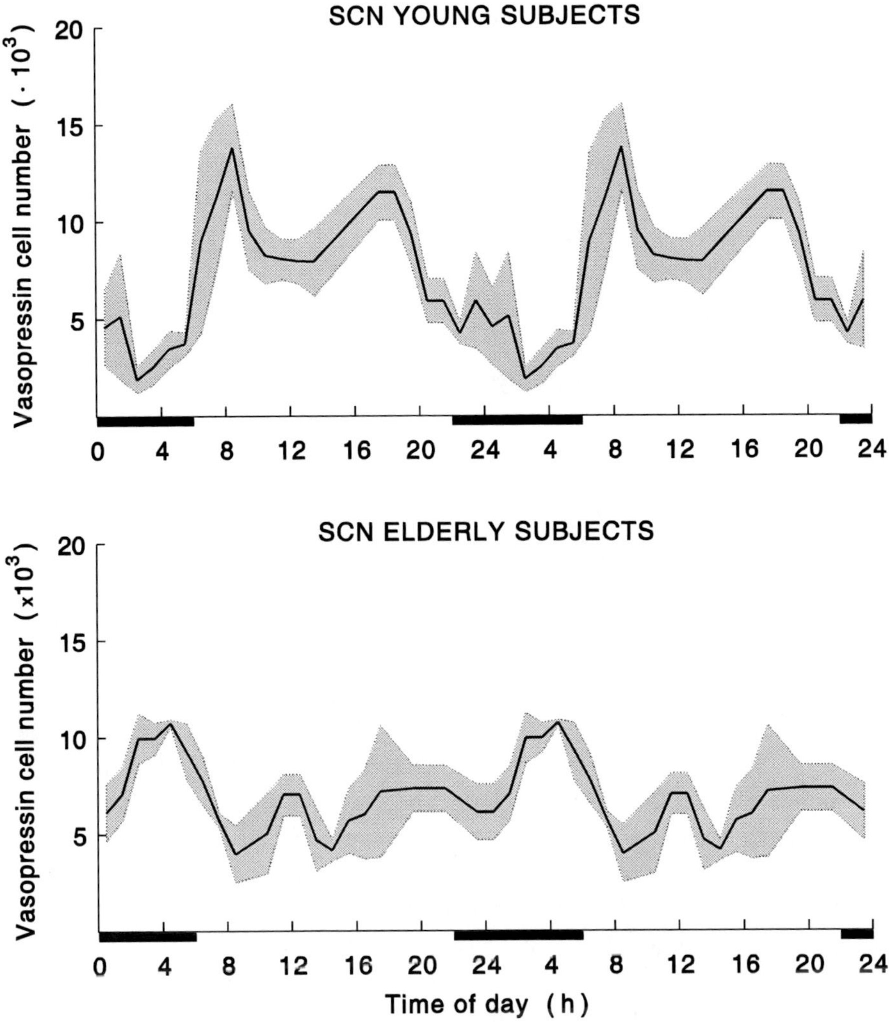

Figure 8. Circadian rhythm in the number of arginine vasopressin (AVP)-expressing neurons in the human suprachiasmatic nucleus (SCN) of (upper panel) young subjects (<50 years of age) and (lower pannel) elderly subjects (>50 years of age). The black bars indicate the night period (22:00-06:00 h). The general trend in the data is enhanced by using a smoothed double plotted curve and is represented by mean ± SEM values. Note the distinct circadian neuropeptide rhythm in the biological clock of young people compared with the disrupted rhythm in older people. Reproduced from Hofman and Swaab (1994).

Immunocytochemical studies, furthermore, provide evidence that degenerative alterations in the human SCN occur at a later phase in life than the reported functional changes in circadian organization. More frequent and prolonged awakenings and shorter sleep periods have already been found in 50- to 60-year-old subjects (Myers and Badia, 1995; Prinz et al., 2000), whereas a reduction in SCN volume and number of AVP-expressing neurons are only present from the age of 80 years onwards (Swaab et al., 1985; 1993; Hofman, 1997). Thus, the observed loss of AVP-expressing neurons in the SCN of very old people may only be a relatively late correlate of functional changes in the biological clock appearing much earlier.

5.4. Pineal Melatonin and Aging

Reduced melatonin concentrations during aging, especially nocturnal levels, have been reported in plasma, cerebro-spinal fluid (CSF) and in urine (reviewed in Skene and Swaab, 2003; Karasek, 2003; Wu and Swaab, 2005). Studies of the major urinary metabolite of melatonin, 6-sulfatoxymelatonin show that age related decrease in melatonin production occurs even as early as 20 – 30 years of age (Kennaway et al., 1999). Zhao and colleagues found that a decline of the nocturnal serum melatonin peak was only significant at the age of 60 and further declined from 70 years of age onwards (Zhao et al., 2002). Even within a fairly narrow age range (40 - 69 years), there is a significant effect of age on the daily excretion of urinary 6-sulfatoxymelatonin (Skene et al., 1990a). Although many reports indicate that melatonin levels decline with age, especially the nocturnal melatonin peak, some recent studies do not support a reduction (Zeitzer et al., 1999; Fourtillan et al., 2001).

Melatonin content in the human pineal has also been found to be reduced with age (Skene et al., 1990b; Wu and Swaab, 2005; Hofman, 2009b; Figure 9). In another study of human postmortem pineal tissue, elderly subjects had lower pineal melatonin contents than younger subjects, but this difference was not statistically significant (Luboshitzky et al., 1998). Besides the age-related decline of melatonin production, age-related changes in the timing of the melatonin rhythm have also been reported (Duffy et al., 2002). Moreover, older subjects enter sleep and awake earlier relative to their nightly melatonin secretory episode, which indicates that aging is also associated with a change in the internal phase relationship between the sleep–wake cycle and the output of the circadian pacemaker.

In addition to the effects of normal aging various studies have shown that melatonin levels are lower in Alzheimer's disease (AD) patients compared with aged matched controls (Skene et al., 1990b; Uchida et al., 1996; Liu et al., 1999;

Mishima et al., 1999; Ohashi et al., 1999; Ferrari et al., 2000; Karasek, 2003). In patients who lack serum melatonin rhythms, clinical symptoms of delirium and sleep-wake disturbance were frequently but not always found (Uchida et al., 1996). In AD patients with disturbed sleep-wake rhythms a higher degree of irregularities in melatonin secretion have been observed (Mishima et al., 1999). Our finding that the daily variations in pineal melatonin and 5-methoxytryptophol content disappeared in AD patients (Skene et al., 1990b; Wu et al., 2003) may be linked with the clinical observations of sleep disorders and sundowning in these patients. Others have reported that there is a selective impairment of the nocturnal melatonin peak in dementia (Ferrari et al., 2000) or that the melatonin levels are increased in AD patients during daytime and that these patients do not react to bright light (Ohashi et al., 1999).

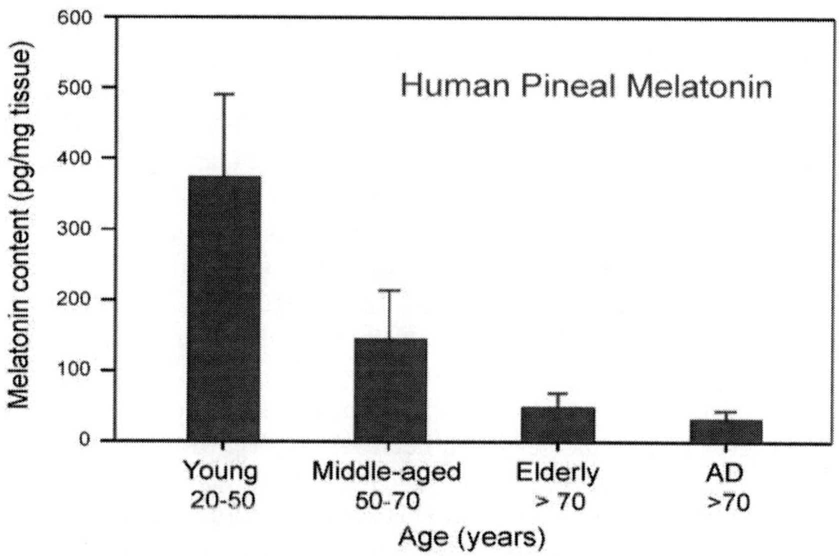

Figure 9. Effect of aging and Alzheimer's disease (AD) on the night-time melatonin level in the human pineal gland. Note the dramatic decline of the nocturnal melatonin production in middle-aged subjects compared to that of young subjects and the further decline after the age of 70 years. Reproduced from Hofman (2009b).

5.5. Sex, Circadian Clock and Aging

In humans, during normal aging, the number of peptidergic neurons in the SCN deteriorate in a sexually dimorphic way (for a review, see Swaab, 1999). The number of VIP-expressing neurons in the SCN of women, for example, did not show any age-related changes as opposed to the neurons expressing AVP, whereas in men a complex pattern of changes was observed of VIP-expressing neurons with advancing age. Between 10 and 40 years, the male SCN contained twice as many VIP neurons as the female, but a subsequent decrease in the number of male VIP neurons between 40 and 65 years of age resulted in fewer VIP neurons in men than in women. After 65 years of age the sex difference remained just short of significance. A significant decrease in the number of VIP-expressing neurons in the SCN was found in female presenile Alzheimer patients only (Zhou et al., 1995).

With respect to the sex differences in age-related changes in VIP neurons it is of interest to mention that there are differences between healthy aged women and men as far as entrained circadian temperature rhythms are concerned that suggest that aging may affect the circadian timing system in a sexually dimorphic way. The acrophase of body temperature was phase-advanced by an average of 1.25 h in the aged women compared to the age-matched men. Women woke up earlier and slept for shorter periods of time (Moe et al., 1991). In addition, the sex differences (Swaab and Hofman, 1995), the presence of sex hormone receptors (Kruijver and Swaab, 2002) and the sexual orientation related differences in the human SCN (Swaab and Hofman, 1990, 1995), reinforce the idea on the possible involvement of this nucleus in sexual behavior or reproduction.

Many findings indicate that changes in the SCN and other parts of the circadian timing system underlie some of the sleep disturbances among elderly people (Dijk et al., 2000; Hofman, 2000; Van Someren, 2000b; Hofman and Swaab, 2006). Based on the hypothesis that increased stimulation of the brain can improve or even restore the decreased neuronal activity (Swaab et al., 2002), it has been demonstrated in aged rats that both overt sleep-wake rhythms (Witting et al., 1993) and cell function in the SCN (Lucassen et al., 1995) can be restored by enhancing the light stimulus the circadian timing system normally uses for synchronization of the sleep-wake rhythm. There is evidence that a similar degree of plasticity exists in the human circadian timing system. Improvement of the sleep-wake rhythm of elderly people has been demonstrated by application of a variety of potent modulators of the circadian timing system, like bright light, melatonin and physical activity (Van Someren et al., 2002; Meijer and Schwartz, 2003). Contrary to treatment with hypnotics, the improvement of sleep following these treatments is without adverse effects and even results in improvement of mood, performance, daytime energy, and quality of life.

CONCLUSION

It is now generally accepted that the suprachiasmatic nucleus (SCN) is the principal neural structure that mediates circadian rhythms in mammals, including man. Consistent with its role in the circadian organization of physiological processes and behavior, recent investigations suggest that measurable biological activities within the SCN, such as glucose utilization, neuronal electrical activity, and protein synthesis exhibit circadian rhythms (Klein et al., 1991; Van Esseveldt et al., 2000).

Animal studies have shown that circadian rhythmicity is disrupted during senescence at various levels of biological organization, including an age-related reduction in the amplitude of the rest-activity cycle (Turek et al., 1995; Weinert, 2000). Analogous to these animal data, age-related alterations have also been described for hormonal rhythms, temperature rhythms, sleep-wakefulness rhythms, as well as for various other physiological and behavioral rhythms in humans (Touitou and Haus, 1992, 2000; Turek et al., 1995; Van Someren, 2000b; Dijk et al., 2000; Van Someren et al., 2002). It appears that aging not only affects the amplitude but also the frequency of circadian rhythms, particularly in dementia (Witting et al., 1990; Turek et al., 1995). Generally, dissociation of internal synchronization of circadian rhythms seems to occur more frequently in the elderly (Touitou and Haus, 1992; 2000). Such age-related changes in the circadian period may have major consequences for the synchronzation of the biological clock with external Zeitgebers as well as for the temporal organization of the multioscillatory system.

The fact that the internal synchronization of various rhythms appears to be disturbed in elderly humans, in addition to alterations in the rhythmicity itself, points towards the circadian oscillator as the site of origin of these age-related changes. The observed reduction in the amplitude of the AVP rhythm of the SCN in elderly people is in accordance with this view, and with findings that the human pineal gland, as part of the circadian timing system, shows similar alterations with an age-related decrease in melatonin synthesis and increased pineal calcium deposits (Skene et al., 1990a; Skene and Swaab, 2003; Wu et al., 2003; Wu and Swaab, 2005). Moreover, the circadian melatonin rhythm becomes increasingly unstable in older people. Similar disorders of the pineal biosynthetic activity have been shown in the aging hamster and gerbil (Reiter, 1995). These findings suggest that degeneration of the SCN-pineal complex could well be the neural substrate for the disrupted circadian rhythms, which have been reported in elderly subjects, demented patients and in depression (Zhou et al., 2001). Experimental studies, for example, show that the free-running period of the rest-activity cycle is directly proportional to the residual volume of the SCN in partly lesioned animals, and such animals show a pronounced reduction in the amplitude and phase advance of daily activity, drinking and eating rhythms (Turek et al., 1995).

In conclusion, the age-related functional changes in circadian organization in humans may be associated with subtle degenerative alterations in the SCN and other parts of the clock, leading to disruptions of the circadian system. In the present review it was shown that this may happen relatively early in life, well before any dramatic neuronal atrophy of the biological clock becomes manifest. Furthermore, the present findings indicate that illuminance and photoperiod may be considered as potential environmental factors controlling the functional activity of the human SCN. If light is the effective Zeitgeber in man, as it is in many organisms, disturbances in processing of light information during aging or in some neurodegenerative and affective diseases may have profound effects on the timing of a variety of physiological and behavioral activities, including sleep (see e.g., Meijer and Schwartz, 2003). In view of the assumed sensitivity of the biological clock to changes in day length and light intensity it would be interesting to investigate whether the synchronizing and antidepressant effects of bright light exposure in sleep disturbances and mood disorders can be improved by applying the therapy in temporal coherence with the circadian cycles underlying the neuronal activity of the biological clock.

REFERENCES

Albrecht, U., Zheng, B., Larkin, D., Sun, Z. S. & Lee, C. C. (2001). *mPer1* and *mPer2* are essential for normal resetting of the cicadian clock. *Journal of Biological Rhythms*, *16*, 100-104.

Antle, M. C. & Silver, R. (2005). Orchestrating time: arrangements of the brain circadian clock. *Trends in Neurosciences*, *28*, 145-151.

Antle, M. C., Foley, D. K., Foley N. C. & Silver R. (2003). Gates and oscillators: a network model of the brain clock. *Journal of Biological Rhythms*, *18*, 339-350.

Asai, M., Yoshinobu, Y., Kaneko, S., Mori, A., Nikaido, T., Moriya, T., Akiyama, M. & Shibata, S. (2001). Circadian expression of *Per* gene mRNA expression in the suprachiasmatic nucleus, paraventricular nucleus, and pineal body of aged rats. *Journal of Neuroscience Research*, *66*, 1133-1139.

Aujard, F., Herzog, E. D. & Block, G. D. (2001). Circadian rhythms in firing rate of individual suprachiasmatic nucleus neurons from adult and middle-aged mice. *Neuroscience*, *106*, 255-261.

Avidan, A. Y. (2005). Sleep in the geriatric patient population. *Seminars in Neurology*, *25*, 52-63.

Bhatt, M. H., Podder, N. & Chokroverty, S. (2005). Sleep and neurodegenerative diseases. *Seminars in Neurology*, *25*, 39-51.

Buijs, R. M. & Kalsbeek, A. (2001). Hypothalamic integration of central and peripheral clocks. *Nature Reviews/ Neuroscience*, *2*, 521-526.

Buijs, R. M., Kalsbeek, A., Romijn, H. J., Pennartz, C. M. A. & Mirmiran, M. (Eds.) (1996). Hypothalamic Integration of Circadian Rhythms. *Progress in Brain Research*, Vol. *111*. Elsevier, Amsterdam.

Cermakian, N. & Sassone-Corsi, P. (2000). Multilevel regulation of the circadian clock. *Nature Reviews/ Molecular Cell Biology*, 1, 59-67.

Claustrat, F., Fournier, I., Geelen, G., Brun, J., Corman, B. & Claustrat, B. (2005). Aging circdian clock gene expression in peripheral tissue in rats. *Pathologie-Biologie (Paris), 53*, 257-260.

Cohen, R. A. & Albers, H. E. (1991). Disruption of the human circadian and cognitive regulation following a discrete hypothalamic lesion: a case study. *Neurology*, *41*, 726-729.

Cornélissen, G., Haus, E. & Halberg, F. (1992). Chronobiologic blood pressure assessment from womb to tomb. In: Touitou, Y., Haus, E. (Eds.). *Biological Rhythms in Clinical and Laboratory Medicine*. Springer Verlag, Berlin, pp. 428-452.

Czeisler, C. A., Dumont, M., Duffy, J. F., Steinberg, J. D., Richardson, G. S., Brown, E. N., Sanchez, R., Rios, C. D. & Ronda, J. M. (1992). Association of sleep-wake habits in older people with changes in output of circadian pacemaker. *Lancet*, *340*, 933-936.

Czeisler, C. A. (1995). The effect of light on the human circadian pacemaker. In: Waterhouse, J. M. (Ed.). *Circadian Clocks and their Adjustment*, Ciba Foundation Symposium 183, John Wiley, Chichester, pp. 254-302.

Daan, S., Albrecht, U., Van der Horst, G. T. J., Illnerová, H., Roenneberg, T., Wehr, T. A. & Schwartz, W. J. (2001). Assembling a clock for all seasons: are there M and E oscillators in the genes? *Journal of Biological Rhythms*, *16*, 105-116.

Dai, J., Swaab, D. F. & Buijs, R. M. (1997). Distribution of vasopressin and vasoactive intestinal polypeptide (VIP) fibers in the human hypothalamus with special emphasis on suprachiasmatic nucleus efferent projections. *Journal of Comparative Neurology*, *383*, 397-414.

Dai, J., Swaab, D. F., Van der Vliet, J. & Buijs, R. M. (1998). Postmortem tracing reveals the organization of hypothalamic projections of the suprachiasmatic nucleus in the human brain. *Journal of Comparative Neurology*, *400*, 87-102.

Dardente, H., Poirel V. J., Klosen, P., Pévet, P. & Masson-Pévet, M. (2002). *Per* and neuropeptide expression in the rat suprachiasmatic nuclei: compartmentalization and differential cellular induction by light. *Brain Research*, *958*, 261-271.

Dijk, D. J., Duffy, J. F. & Czeisler, C. A. (2000). Contribution of circadian physiology and sleep homeostasis to age-related changes in human sleep. *Chronobiology International*, *17*, 285-311.

Duffy, J. F., Zeitzer, J. M., Rimmer, D. W., Klerman, D. B., Dijk, D. J. & Czeisler, C. A. (2002). Peak of circadian melatonin rhythm occurs later within the sleep of older subjects. *American Journal of Physiology. Endocrinology and Metabolism*, *282*, E297-E303.

Dunlap J. C. (1999). Molecular bases for circadian clocks. *Cell*, *96*, 271-290.

Eastman, C. I., Mistlberger, R. E. & Rechtschaffen, A. (1984). Suprachiasmatic nuclei lesions eliminate circadian temperature and sleep rhythms in the rat. *Physiology and Behavior*, *32*, 357-368.

Ferrari E., Arcaini, A., Gornati, R., Pelanconi, L., Cravello, L., Fioravanti, M., Solerte, M. B. & Maqri, F. (2000). Pineal and pituitary-adrenocortical function in physiological aging and in senile dementia. *Experimental Gerontology*, *35*, 1239-1250.

Fourtillan, J. B., Brisson, A. M., Fourtillan, M., Ingrand, L., Decourt, J. P. & Girault, J. (2001). Melatonin secretion occurs at a constant rate in both young and older men and women. *American Journal of Physiology. Endocrinology and Metabolism*, *280*, E11-E22.

Hamada T., Antle M. C. & Silver, R. (2004). Temporal and spatial expression patterns of canonical clock genes and clock-controlled genes in the suprachiasmatic nucleus. *European Journal of Neuroscience*,*19*, 1741-1748.

Hamada, T., LeSauter J., Venuti J. M. & Silver R. (2001). Expression of *Period* genes: rhythmic and nonrhythmic compartments of the suprachiasmatic nucleus pacemaker. *Journal of Neuroscience*, *21*, 7742-7750.

Hannibal, J. (2002). Neurotransmitters of the retino-hypothalamic tract. *Cell and Tissue Research*, *309*, 73-88.

Hastings, M. H. & Herzog, E. D. (2004). Clock genes, oscillators, and cellular networks in the suprachiasmatic nuclei. *Journal of Biological Rhythms*, *19*, 400-413.

Hastings, M. H. & Maywood, E. S. (2000). Circadian clocks in the mammalian brain. *BioEssay*, *22*, 23-34.

Hastings, M. H., Reddy, A. B. & Maywood, E. S. (2003). A clockwork web: circadian timing in brain and periphery, in health and disease. *Nature Reviews/ Neuroscience*, *4*, 449-461.

Hofman, M. A. (1997). Lifespan changes in the human hypothalamus. *Experimental Gerontology*, *32*, 559-575.

Hofman, M. A. (2000). The human circadian clock and aging. *Chronobiology International*, *17*, 245-259.

Hofman, M. A. (2003). Circadian oscillations of neuropeptide expression in the human biological clock. *Journal of Comparative Physiology* A, *189*, 823-831.

Hofman, M. A. (2004). The brain's calendar: neural mechanisms of seasonal timing. *Biological Reviews*, *79*, 61-77.

Hofman, M.A. (2009a). Seasonal timing: neural mechanisms. In: Squire, L.R. (Ed.). *Encyclopedia of Neuroscience.* Academic Press, Oxford, pp. 509-515.

Hofman, M. A. (2009b). Human circadian timing system. In: Binder, M.D., Hirokawa, N., Windhorst, U. (Eds.). *Encyclopedia of Neuroscience.* Springer Verlag, Heidelberg, pp. 1869-1873.

Hofman, M. A., Fliers, E., Goudsmit, E. & Swaab, D. F. (1988). Morphometric analysis of the suprachiasmatic and paraventricular nuclei in the human brain: sex differences and age-dependent changes. *Journal of Anatomy*, *160*, 127-143.

Hofman, M. A. & Swaab, D. F. (1992). The human hypothalamus: comparative morphometry and photoperiodic influences. In: Swaab, D. F., Hofman, M. A., Mirmiran, M., Ravid, R., Van Leeuwen, F. W. (Eds.). The Human Hypothalamus in Health and Disease. *Progress in Brain Research*, Vol. *93*. Elsevier, Amsterdam, pp. 133-149.

Hofman, M. A. & Swaab, D. F. (1993). Diurnal and seasonal rhythms of neuronal activity in the suprachiasmatic nucleus of humans. *Journal of Biological Rhythms*, *8*, 283-295.

Hofman, M. A. & Swaab, D. F. (1994). Alterations in circadian rhythmicity of the vasopressin-producing neurons of the human suprachiasmatic nucleus (SCN) with aging. *Brain Research*, *651*, 134-142.

Hofman, M. A. & Swaab, D. F. (1995). Influence of aging on the seasonal rhythm of the vasopressin-expressing neurons in the human suprachiasmatic nucleus. *Neurobiology of Aging*, *116*, 965-971.

Hofman, M. A. & Swaab, D. F. (2002). A brain for all seasons: cellular and molecular mechanisms of photoperiodic plasticity. In: Hofman, M. A., Boer, G. J., Holtmaat, A. J.G. D., Van Someren, E. J.W., Verhaagen, J., Swaab, D. F. (Eds.). Plasticity in the Adult Brain: From Genes to Neurotherapy. *Progress in Brain Research*, Vol. 138. Elsevier, Amsterdam, pp. 255-280.

Hofman, M. A. & Swaab, D. F. (2006). Living by the clock: The circadian pacemaker in older people. *Ageing Research Reviews*, *5*, 33-51.

Hofman, M. A., Zhou, J. N. & Swaab, D. F. (1996). Suprachiasmatic nucleus of the human brain: an immunocytochemical and morphometric analysis. *Anatomical Record*, *244*, 552-562.

Jagota, A., De la Iglesia, H. O. & Schwartz, W. J. (2000). Morning and evening circadian oscillations in the suprachiasmatic nucleus *in vitro*. *Nature Neuroscience*, *3*,372-376.

Kalsbeek, A. & Buijs, R. M. (2002). Output pathways of the mammalian suprachiasmatic nucleus: coding circadian time by transmitter selection and specific targeting. *Cell and Tissue Research*, *309*, 109-118.

Kalsbeek, A., Palm, I. F., La Fleur, S. E., Scheer, F. A., Perreau-Lenz, S., Ruiter, M., Kreier, F., Cailotto, C. & Buijs, R. M. (2006). SCN outputs and the hypothalamic balance of life. *Journal of Biological Rhythms*, *21*, 458-469.

Karasek, M. (2003). Melatonin in human aging and age-related diseases. *Experimental Gerontology*, *39*, 1723-1729.

Karatsoreos, I. N., Yan, L., LeSauter, J. & Silver, R. (2004). Phenotypic matters: identification of light-responsive cells in the mouse suprachiasmatic nucleus. *Journal of Neuroscience*, *24*, 68-75.

Kawamoto, K., Nagano, M., Kanda, F., Chihara, K., Shigeyoshi, Y. & Okamura, H. (2003). Two types of VIP neuronal components in rat suprachiasmatic nucleus. *Journal of Neuroscience Research*, *74*, 852-857.

Kenneway, D. J., Lushington, K., Dawson, D., Lack, L., Van den Heuvel, C. & Rogers, N. (1999). Urinary 6-sulfatoxymelatonin secretion ang aging: new results and a critical review of the literature. *Journal of Pineal Research*, *27*, 210-220.

King, D.P. & Takahashi, J.S. (2000). Molecular genetics of circadian rhythms in mammals. *Annual Review of Neuroscience*, *23*, 713-742.

King, V. M., Chahad-Ehlers, S., Shen, S., Harmar, A. J., Maywood, E. S. & Hastings, M. H. (2003). A hVIPR transgene as a novel tool for the analysis of circadian function in the mouse suprachiasmatic nucleus. *European Journal of Neuroscience*, *17*, 822-832.

Klein, D. C., Moore, R. Y. & Reppert, S. M. (Eds.) (1991). *Suprachiasmatic Nucleus; The Mind's Clock*. Oxford University Press, New York.

Kolker, D. E., Fukuyama, H., Huang, D. S., Takahashi, J. S., Horton, T. H. & Turek, F. W. (2003). Aging alters circadian and light-induced expression of clock genes in golden hamsters. *Journal of Biological Rhythms*, *18*, 159-169.

Kolker, D. E., Vitaterna, M. H., Fruechte, E. M., Takahashi, J. S. & Turek, F. W. (2004). Effects of age on circadian rhythms are similar in wild-type and heterozygous *Clock* mutant mice. *Neurobiology of Aging*, *25*, 517-523.

Kruijver, F. P. M. & Swaab, D. F. (2002). Sex hormone receptors are present in the human suprachiasmatic nucleus. *Neuroendocrinology*, *75*, 296-305.

Leak, R. K. & Moore, R. Y. (2001). Topographic organization of suprachiasmatic nucleus projection neurons. *Journal of Comparative Neurology*, *433*, 312-314.

Liu, R. Y., Zhou, J. N., Van Heerikhuize, J., Hofman, M. A. & Swaab, D.F . (1999). Decrease melatonin levels in postmortem cerebrospinal fluid in relation to aging, Alzheimer's disease and apolipoprotein E *epsilon*-4/4 genotype. *Journal of Clinical Endocrinology and Metabolism, 84*, 323-327.

Liu, R. Y., Zhou, J. N., Hoogendijk, W. J. G., Van Heerikhuize, J., Kamphorst, W., Unmehopa, U. A., Hofman, M. A. & Swaab, D. F. (2000). Decreased vasopressin gene expression in the biological clock of Alzheimer's disease patients with and without depression. *Journal of Neuropathology and Experimental Neurology*, *59*, 314-322.

Luboshitzky, R., Yanai, D., Shen-Orr, Z., Israeli, E., Herer, P. & Lavie, P. (1998). Daily and seasonal variations in the concentration of melatonin in the human pineal gland. *Brain Research Bulletin*, *47*, 271-276.

Lucassen, P. J., Hofman, M. A. & Swaab, D. F. (1995). Increased light intensity prevents the age-related losss of vasopressin-expressing neurons in the rat suprachiasmatic nucleus. *Brain Research*, *693*, 261–266.

Mai, J. K., Kedziora, O., Teckhaus, L. & Sofroniew, M. V. (1991). Evidence for subdivisions in the human suprachiasmatic nucleus. *Journal of Comparative Neurology*, *305*, 508-525.

Meijer, J. H., Rusak, B. & Ganshirt, G. (1992). The relation between light-induced discharge in the suprachiasmatic nucleus and phase shifts of hamster circadian rhythms. *Brain Research*, *598*, 257-63.

Meijer, J. H. & Schwartz, W. J. (2003). In search of the pathways for light-induced pacemaker resetting in the suprachiasmatic nucleus. *Journal of Biological Rhythms*, *18*, 235-249.

Mishima, K., Okawa, M., Shimizu, T. & Hishikawa, Y. (2001). Diminished melatonin secretation in the elderly caused by insufficient environmental illumination. *Journal of Clinical Endocrinology and Metabolism*, *86*, 129-134.

Moe, K. E., Vitiello, M. V., Larsen, L. H. & Prinz, P. N. (1995). Cognitive processes and sleep disturbances: Sleep–wake patterns in Alzheimer's disease: relationships with cognition and function. *Journal of Sleep Research*, *4*, 15–20.

Monk, T. Y. & Kupfer, D. J. (2000). Circadian rhythms in healthy aging – effects downstream from the pacemaker. *Chronobiology International*, *17*, 355-368.

Moore, R. Y. (1991). Disorders of circadian function and the human circadian timing system. In: Klein, D.C, Moore, R.Y., Reppert, S.M., (Eds.). *Suprachiasmatic Nucleus: The Mind's Clock.* Oxford University Press, New York, pp. 429-441.

Moore, R. Y. (1992). The organization of the human circadian timing system. In: Swaab, D. F., Hofman, M. A., Mirmiran, M., Ravid, R., Van Leeuwen, F.W. (Eds.). The Human Hypothalamus in Health and Disease. *Progress in Brain Research*, Vol. *93*. Elsevier, Amsterdam, pp. 101-117.

Moore, R. Y. & Silver, R. (1998). Suprachiasmatic nucleus organization. *Chronobiology International*, *15*, 475-487.

Moore, R. Y., Speh, J. C. & Leak, R. K. (2002). Suprachiasmatic nucleus organization. *Cell and Tissue Research*, *309*, 89-98.

Morin, L.P. (1988). Age-related changes in hamster circadian period, entrainment, and rhythm splitting. *Journal of Biological Rhythms*, *3*, 237-248.

Myers, B. L. & Badia, P. (1995). Changes in circadian rhythms and sleep quality with aging: Mechanisms and interventions. *Neuroscience and Biobehavioral Reviews*, *19*, 553-571.

Nakamura, W., Honma, S., Shirakawa, T. & Honma, K. (2001). Regional pacemakers composed of multiple oscillator neurons in the rat suprachiasmatic nucleus. *European Journal of Neuroscience*, 14, 666-674.

Ohashi, Y., Okamoto, N., Uchida, K., Iyo, M., Mori, N. & Morita, Y. (1999). Daily rhythm of serum melatonin levels and effect of light exposure in patients with dementia of the Alzheimer's type. *Biological Psychiatry*, *45*, 1646-1652.

Panda, S., Hogenesch, J. B. & Kay, S. A. (2002). Circadian rhythms from flies to human. *Nature*, *417*, 329-335.

Prinz, P. N., Bailey, S. L. & Woods, D. L. (2000). Sleep impairments in healty seniors: roles of stress, cortisol, and interleukin-1 beta. *Chronobiology International*, *17*, 391-404.

Quintero, J. E., Kuhlman, S. J. & McMahon, D. G. (2003). The biological clock nucleus: a multiphasic oscillator network regulated by light. *Journal of Neuroscience*, *23*, 8070-8076.

Reiter, R. J. (1995). The pineal gland and melatonin in relation to aging: summary of the theories and of the data. *Experimental Gerontology*, *30*, 199-212.

Renfrew, J. W., Pettigrew, K. D. & Rapoport, S. I. (1987). Motor activity and sleep duration as a function of age in healthy men. *Physiology and Behavior*, *41*, 627-634.

Reppert, S. M. & Weaver, D. R. (2001). Molecular analysis of mammalian circadian rhythms. *Annual Reviews of Physiology*, *63*, 647-676.

Reppert, S. M. & Weaver, D. R. (2002). Coordination of circadian timing in mammals. *Nature*, *418*, 935-941.

Reuss, S. (1996). Components and connections of the circadian timing system in mammals. *Cell and Tissue Research*, *285*, 353-378.

Romijn, H. J., Sluiter, A. A., Pool, C. W., Wortel, J. & Buijs, R. M. (1996). Differences in colocalization between Fos and PHI, GRP, VIP, and VP in neurons of the rat suprachiasmatic nucleus after a light stimulus during the phase-delay versus the phase-advance period of the night. *Journal of Comparative Neurology*, *372*, 1-8.

Romijn, H. J., Van Uum, J. F., Emmering, J., Goncharuk, V. & Buijs, R. M. (1999). Colocalization of VIP and AVP in neurons of the paraventricular, supraoptic and suprachiasmatic nucleus. *Brain Research*, *832*, 47-53.

Sack, R. L. & Lewy, A. J. (2001). Circadian rhythm sleep disorders: lessons from the blind. *Sleep Medicine Reviews*, *5*, 189-206.

Sadun, A. A., Schaechter, J. D. & Smith, L. E. H. (1993). A retinohypothalamic pathway in man: light mediation of circadian rhythms. *Brain Research*, *302*, 371-377.

Saper, C. B., Lu, J., Chou, T. C. & Gooley, J. (2005). The hypothalamic integrator for circadian rhythms. *Trends in Neurosciences*, *28*, 152-157.

Satinoff, E., Li, H., Tcheng, T. K., Liu, C., McArthur, A. J., Medanic, M. & Gillette, M. U. (1993). Do the suprachiasmatic nuclei oscillate in old rats as they do in young ones? *American Journal of Physiology*, *265*, R1216-R1222.

Schwartz, W. J., Bosis, N. A. & Hedley-Whyte, E. T. (1986). A discrete lesion of ventral hypothalamus and optic chiasm that disturbed the daily temperature rhythm. *Journal of Neurology*, *233*, 1-4.

Schwartz, W. J., Carpino, Jr, A., De la Iglesia, H. O., Baler, R., Klein, D. C., Nakabeppu, Y. & Aronin, N. (2000). Differential regulation of *fos* family genes in the ventrolateral and dorsomedial subdivisions of the rat suprachiasmatic nucleus. *Neuroscience*, *98*: 535-547.

Shirakawa, T., Honma, S. & Honma, K. (2001). Multiple oscillators in the suprachiasmatic nucleus. *Chronobiology International*, *18*, 371-387.

Skene, D. & Swaab, D. F. (2003). Melatonin rhythmicity: effect of age and Alzheimer's disease. *Experimental Gerontology*, *38*, 199-206.

Skene, D. J., Bojkowski, C. J., Currie, J. E., Wright, J., Boulter, P. S. & Arendt, J. (1990a). 6-sulphatoxymelatonin production in breast cancer patients. *Journal of Pineal Research*, *8*, 269-276.

Skene, D. J, Vivien-Roels, B., Sparks, D. L., Hunsaker, J. C., Pevet, P., Ravid, R. & Swaab, D. F. (1990b). Daily variation in the concentrations of melatonin and 5-methoxytryptophol in the human pineal gland: effect of age and Alzheimer's disease. *Brain Research*, *528*, 170-174.

Spoelstra, K., Albrecht, U., Van der Horst, G. T., Brauer, V. & Daan, S. (2001). Phase responses to light pulses in mice lacking functional *Per* or *Cry* genes. *Journal of Biological Rhythms*, *19*, 518-529.

Stopa, E. G., Volicer, L., Kuo-Leblanc, V., Harper, D., Lathi, D., Tate, B. & Satlin, A. (1999). Pathologic evaluation of the human suprachiasmatic nucleus in severe dementia. *Journal of Neuropathology and Experimental Neurology*, *58*, 29-39.

Sutin, E. L., Dement, W. C., Heller, H. C. & Kilduff, T. S. (1993). Light-induced gene expression in the suprachiasmatic nucleus of young and aging rats. *Neurobiology of Aging*, *14*, 441-446.

Swaab, D. F. (1999). Biological rhythms in health and disease: the suprachiasmatic nucleus and the autonomic nervous system. In: Vinken, P. G., Bruyn, G. W., Appenzeller, O. (Eds.). *Handbook of Clinical Neurology*, Vol. *74*. Elsevier, Amsterdam, pp. 467–521.

Swaab, D. F. (2004). The Human Hypothalamus: Basic and Clinical Aspects. Part II: Neuropathology of the Human Hypothalamus and Adjacent Brain Structures. *Handbook of Clinical Neurology*, vol. 80, Elsevier, Amsterdam.

Swaab, D. F. & Hofman, M. A. (1990). An enlarged suprachiasmatic nucleus in homosexual men. *Brain Research*, *537*, 141-148.

Swaab, D. F. & Hofman, M. A. (1995). Sexual differentiation of the human hypothalamus in relation to gender and sexual orientation. *Trends in Neurosciences*, *18*, 264-270.

Swaab, D. F., Fliers, E. & Partiman, T. S. (1985). The suprachiasmatic nucleus of the human brain in relation to sex, age and senile dementia. *Brain Research*, *342*, 37-44.

Swaab, D. F., Dubelaar, E. J., Hofman, M. A., Scherder, E. J. A., Van Someren, E. J. W. & Verwer, R. W. H. (2002). Brain aging and Alzheimer's disease; use it or lose it. In: Hofman, M. A., Boer, G. J., Holtmaat, A. J. G. D. Van Someren, E. J. W., Verhaagen, J., Swaab, D. F. (Eds.). Plasticity in the Adult Brain: From Genes to Neurotherapy. *Progress in Brain Research,* Vol. 138. Elsevier, Amsterdam, pp. 343-373.

Swaab, D. F., Grundke-Iqbal, I., Iqbal, K., Kremer, H. P., Ravid, R. & Van de Nes, J. A. (1992). Tau and ubiquitin in the human hypothalamus in aging and Alzheimer's disease. *Brain Research*, *590*, 239–249.

Swaab, D. F., Hofman, M. A., Lucassen, P. J., Purba, J. S., Raadsheer, F. C. & Van de Nes, J.A. (1993). Functional neuroanatomy and neuropathology of the human hypothalamus. *Anatomy and Embryology* (Berlin), *187*, 317-30.

Swaab, D. F., Van Someren, E. J. W., Zhou, J. N. & Hofman, M. A. (1996). Biological rhythms in the human life cycle and their relationship to functional changes in the suprachiasmatic nucleus. In: Buijs, R.M., Kalsbeek, A., Romijn, H. J, Pennartz, C. M. A., Mirmiran, M. (Eds.). Hypothalamic Integration of Circadian Rhythms. *Progress in Brain Research*, Vol. 111. Elsevier, Amsterdam, pp. 349-368.

Touitou, Y. & Haus, E. (1992). Biological rhythms and aging. In: Touitou, Y., Haus, E., (Eds.). *Biological Rhythms in Clinical and Laboratory Medicine*. Springer, Berlin, pp. 188-207.

Touitou, Y. & Haus, E. (2000). Alterations with aging of the endocrine and neuroendocrine circadian system in humans. *Chronobiology International*, *17*, 369-390.

Turek, F. W., Penev, P., Zhang, Y., Van Reeth, O. & Zee, P. (1995). Effects of age on the circadian system. *Neuroscience and Biobehavioral Reviews*, *19*, 53-58.

Uchida, K., Okamoto, N., Ohara, K. & Morita, Y. (1996). Daily rhythms of serum melatonin in patients with dementia of degenerate type. *Brain Research*, *717*, 154-159.

Van de Nes, J. A., Kamphorst, W., Ravid, R. & Swaab, D. F. (1998). Comparison of beta-protein/A4 deposits and Alz-50-stained cytoskeletal changes in the hypothalamus and adjoining areas of Alzheimer's disease patients: amorphic plaques and cytoskeletal changes occur independently. *Acta Neuropathologica* (Berlin), *96*, 129–138.

Van Esseveldt, K. E., Lehman, M. N. & Boer, G. J. (2000). The suprachiasmatic nucleus and the circadian time-keeping system revisited. *Brain Research Reviews*, *33*, 44-77.

Van Someren, E. J. W. (Ed.) (2000a). Circadian Rhythms and Sleep in Human Aging. *Chronobiology International* (Special Issue), Vol. *17*.

Van Someren, E. J. W. (2000b). More than a marker: interaction between the circadian regulation of temperature and sleep, age-related changes and treatment possibilities. *Chronobiology International*, *17*, 313-354.

Van Someren, E. J. W., Hagebeuk, E. E. O., Lijzenga, C., Scheltens, P., De Rooij, S. E., Jonker, C., Pot, A. M. & Mirmiram, M. (1996). Circadian rest-activity rhythm disturbances in Alzheimer's disease. *Biological Psychiatry*, *40*, 259-270.

Van Someren, E. J. W., Riemersma, R. F. & Swaab, D. F. (2002). Functional plasticity of the circadian timing system in old age: light exposure. In: Hofman, M. A., Boer, G. J., Holtmaat, A. J. G. D., Van Someren, E. J. W., Verhaagen, J., Swaab, D. F. (Eds.). Plasticity in the Adult Brain: From Genes to Neurotherapy. *Progress in Brain Research, Vol. 138*. Elsevier, Amsterdam, pp. 205-231.

Vitiello, M. V. & Prinz, P. N. (1990). Sleep and sleep disorders in normal aging. In: Thorpy, M. J., (Ed.). *Handbook of Sleep Disorders, Neurological Diseases and Therapy*. Marcel Dekker, New York, pp. 139-151.

Wakamatsu, H., Takahashi, S., Moriya, T., Inouye, S. T., Okamura, H., Akiyama, M. & Shibata, S. (2001). Additive effect of mPer1 and mPer2 antisense oligonucleotides on light-induced phase shift. *NeuroReport*, *12*, 127-131.

Wehr, T. A. (2001). Photoperiodism in humans and other primates: evidence and implications. *Journal of Biological Rhythms*, *16*, 348-364.

Weinert, D. (2000). Age-dependent changes of the circadian system. *Chronobiology International*, *17*, 261-284.

Weinert, H., Weinert, D., Schurov, I., Maywood, E. S. & Hastings, M. H. (2001). Impaired expression of the *mPer2* circadian clock gene in the suprachiasmatic nuclei of aging mice. *Chronobiology International*,18, 559-565.

Weitzman, E. D., Moline, M. L., Czeisler, C. A. & Zimmerman, J. C. (1982). Chronobiology of aging: temperature, sleep-wake rhythms and entrainment. *Neurobiology of Aging*, *3*, 299-309.

Witting, W., Kwa, J. H., Eikelenboom, P., Mirmiran, M. & Swaab, D. F. (1990). Alterations in the circadian rest-activity rhythm in aging and Alzheimer's disease. *Biological Psychiatry*, *27*, 563-572.

Witting, W., Mirmiran, M., Bos, N. P. & Swaab, D. F. (1993). Effect of light intensity on diurnal sleep-wake distribution in young and old rats. *Brain Research Bulletin*, *30*, 157-162.

Wu, Y. H., Feenstra, M. G. P., Zhou, J. N., Liu, R. Y., Toranó, J. S., Van Kan, H. J. M., Fischer, F., Ravid, R. & Swaab, D. F. (2003). Molecular changes underlying reduced pineal melatonin levels in Alzheimer's disease: alterations in preclinical and clinical stages. *Journal of Clinical Endocrinology and Metabolism*, *88*, 5898-5906.

Wu, Y. H. & Swaab, D. F. (2005). The human pineal gland and melatonin in aging and Alzheimer's disease. *Journal of Pineal Research*, *38*, 145-152.

Yamaguchi, S., Isejima, H., Matsuo, T., Okura, R., Yagita, K., Kobayashi, M. & Okamura, H. (2003). Synchronization of cellular clocks in the suprachiasmatic nucleus. *Science, 302*, 1408-1412.

Yamazaki, S., Straume, M., Tei, H., Sakaki, Y., Menaker, M. & Block, G. D. (2002). Effects of aging on central and periheral mammalian clocks. *Proceedings of the National Academy of Science USA*, 99, 10801-10806.

Yan, L. & Okamura, H. (2002). Gradients in the circadian expression of *Per1* and *Per2* genes in the rat surpachiasmatic nucleus. *European Journal of Neuroscience*, *15*, 1153-1162.

Yan, L. & Silver, R. (2002). Differential induction and localization of *mPer1* and *mPer2* during advancing and delaying phase shifts. *European Journal of Neuroscience*, *16*, 1531-1549.

Yan, L. & Silver, R. (2004). Resetting the brain clock: time-course and localization of mPER1 and mPER2 protein expression in suprachiasmatic nuclei during phase shifts. *European Journal of Neuroscience*, *19*, 1105-1109.

Yan, L., Takekida, S., Shigeyoshi, Y. & Okamura, H. (1999). *Per1* and *Per2* gene expression in the rat suprachiasmatic nucleus; circadian profile and the compartment-specific response to light. *Neuroscience, 94*, 141-150.

Zeitzer, J. M., Daniels. J. E., Duffy, J. F., Klerman, E. B., Shanahan, T. L., Dijk, D. J. & Czeisler, C. A. (1999). Do plasma melatonin concentrations decline with age? *American Journal of Medicine*, *107*, 432-436.

Zhao, Z. Y., Xie, Y., Fu, Y. R., Bogdan, A. & Touitou, Y. (2002). Aging and the circadian rhythm of melatonin: a cross-sectional study of Chinese subjects 30-110 yr of age. *Chronobiology International*, *19*, 1171-1182.

Zhou, J. N., Hofman, M. A. & Swaab, D. F. (1995). VIP neurons in the human SCN in relation to sex, age, and Alzheimer's disease. *Neurobiology of Aging*, *16*, 571-576.

Zhou, J. N., Riemersma, R., Unmehopa, U. A., Hoogendijk, W. J., Van Heerikhuize, J. J., Hofman, M. A. & Swaab, D. F. (2001). Alterations in arginine vasopressin neurons in the suprachiasmatic nucleus in depression. *Archives of General Psychiatry*, *58*, 655-662.

In: Biological Clocks: Effects on Behavior... ISBN: 978-1-60741-251-9
Editors: O. Salvenmoser et al. pp. 39-72

Chapter 2

PHOTOSENSITIVITY A DISREGARDED ATTRIBUTE TO ANALYZE "PHOTOPERIODIC CLOCKS"

Hubert R. Spieth* and Katharina Strauß
University of Bielefeld, Department of Animal Ecology, Morgenbreede 45, D-33615 Bielefeld

ABSTRACT

To interpret seasonal timing in arthropods the change of photosensitivity to light/dark cycles during certain ontogenetic phases has been more or less ignored when modelling photoperiodic responses. The consequences of changing photosensitivity are exemplified using the large white butterfly, *Pieris brassicae*, and its response to daylength. It is an essential character in two ways: 1) Because of the 'required day number' (RDN) necessary to induce a specific response, which changes with photosensitivity. 2) Because of perception of increasing and decreasing daylength, which is not a specific ability of a species but a consequence of changing photosensitivity.

With photosensitivity-tests, evidence for a general concept of quantitative time measurement in arthropods is presented. In this context the findings reveal a converse effect of the same photoperiod depending on the light/dark regime in which an individual grows up. This opposite effect supports the argument for the existence of two independent targets for light/dark-cycles, interpreted as two antagonistic time measurement systems,

* Correspondence Author: Phone: +49 521 1062701 E-mail : hubert.spieth@uni-bielefeld.de

and gives evidence for a 'double circadian oscillator clock' mechanism which is based on two submechanisms, a 'short-night determining system' and a separate 'long-night determining system'. The existence and independence of two systems is shown by differences in long-night vs. short-night responses regarding photosensitivity, temperature dependence, duration of diapause, and heritable factors.

QUANTITATIVE VS. QUALITATIVE TIME MEASUREMENT

This chapter is not on genetic or molecular mechanisms of biological clocks but rather on recent findings that support a model for seasonal rhythms comprising two independent measuring systems. The existence of two systems for the regulation of photoperiodic responses is an important factor when interpreting genetic or molecular results. For example, molecular biologists must keep in mind that patterns of genetic activity are not necessarily limited to one photoperiodic measuring system but may possibly be allocated in two systems.

Research on the biological clocks of invertebrates has a long tradition. Based on previous knowledge of photoperiodicity in botany, crucial advances in the research of daily rhythms of animals were made in the 1950s (see Lees 1955; Bünning 1958; Danilevskii 1965). Since then, countless studies on the chronobiology of arthropods, especially insects, have been conducted (Beck 1980; Tauber et al. 1986; Danks 1987; Zaslavski 1988; Saunders 2002). Of these, several have focused on elucidating rhythmic processes in biological systems. Intensive efforts are therefore currently aimed at unraveling the physiologically complex mechanisms used by arthropods and other organisms to measure time (Danks 2003).

Relatively little was known about the molecular processes of biological clocks until the discovery in the 1990s of the now well-known "clock genes" (Saunders et al. 2004), but even then knowledge was mainly restricted to theoretical descriptions of photoperiodic responses. In the absence of insight into the molecular processes actually taking place within the studied system, more than a dozen different models describing photoperiodic responses were developed (VazNunes & Saunders 1999; Saunders 2002). While there was a consensus on the necessity of a light/dark switch for photoperiodic time measurement, the models differed in their measuring mechanisms. Some photoperiodic responses are best described as a "self-sustained oscillation," others as an "hourglass clock" in which a light pulse is needed after each measuring cycle to induce the next cycle. Generally, most models are based on a combination of these measuring principles.

The fact that so many different models have been developed for a single process indicates the problem of explaining all observed photoperiodic responses with one universal mechanism. Also, it was doubtful whether the methods used to detect circadian-based clocks (Nanda-Hamner and Bünsow protocol) were actually capable of confirming their existence (Dumortier & Brunnarius 1987; Dumortier 1994; Veerman 1994, 2001). Subsequently, the time-measuring mechanisms were considered to be of polyphyletic origin, which validated different interpretations of the clock principle in different species (Vaz Nunes & Saunders 1999; Danks 2003; Saunders 2005). However, the goal to find the one and only all-explaining model never ceased and, until today, researchers continue to develop increasingly sophisticated models (Zaslavski 1996; Veerman 2001; Bradshaw et al. 2003).

A large step towards this goal was made by attributing the apparently different measuring principles of oscillator vs. hourglass to the same measuring process. By depicting oscillation as a time-delayed feedback mechanism, it was possible to create dampened oscillations that complied with the attributes of the hourglass principle (Lewis & Saunders 1987; Saunders & Lewis 1987; Saunders 2005). The novelty was that hourglass- or oscillator-based measuring systems no longer had to be attributed to different mechanisms but could be based on the same physiological process, one that could lead to either a self-sustained oscillator or an hourglass result via differentiated regulation.

Another prevailing question is whether time-measurement for seasonal timing is a qualitative or a quantitative process. Long-day conditions can lead to a reaction in plants and animals that is suppressed under short-day conditions, and vice versa. If we estimate a critical photoperiod for reactions like these, then if this critical photoperiod is exceeded a long-day response is induced, and if not, a short-day response ensues. The most basic assumption from this observation is that time-keeping is a process with a robust threshold that qualitatively determines whether a certain light/dark cycle is a short or a long day. Most models of photoperiodic time measurement are based on such a qualitative mechanism. Only through the studies of Kimura (1990) on *Drosophila*, Hardie (1990) on vetch aphids, and Spieth & Sauer (1991) on the large white butterfly has it become increasingly evident that measuring light/dark cycles is a quantitative process.

A pioneer in quantitative time measurement was Zaslavski (1972). His studies on several insects and mites (Zaslavski & Fomenko 1980, 1983) laid the foundation for the first concept of quantitative time measurement mechanisms (1988), the “commanding link model.”

Nevertheless results indicating a threshold-based response have been consistently reported (Kroon et al. 1997; Hua et al. 2005). The ongoing discussion on whether time measurement is qualitative or quantitative is, among other things, related to the fact that it could be rather difficult to present evidence for a quantitative mechanism. In order to prove its existence experimentally, it is absolutely crucial to have a close look at two phases of the photoperiodic response of the tested population. The first phase is the photoperiod with the longest possible scotophase leading to non-diapause responses in all individuals, classically the "shortest long day." The second phase is the photoperiod with the shortest possible scotophase inducing diapause in all individuals, thus the "longest short day." The most robust results for a quantitative measurement are gained by photoperiodic backgrounds close to these thresholds. These backgrounds prevent individuals from going into very deep diapause (at "short short days") or very deep non-diapause (at "long long days") (Gibbs 1975). If this is disregarded, these backgrounds can induce a very consolidated developmental condition, one that might not be reverted by any antagonistic photoperiod (Zaslavski 1995). Thus, it becomes impossible to detect quantitative measurement.

Furthermore, it is important to avoid using photoperiods which belong to the transition phase from diapause to non-diapause responses, as it is then very difficult to interpret results on a reliable basis (Volkovich & Goryshin 1982; Volkovich 1987). This issue is addressed in greater detail later in this chapter.

Despite the different ideas on the function of the photoperiodic clock, the assumption that the clock is based on just one light/dark measuring system was undisputed for many years. This has changed, too. First, one has to distinguish strictly between circadian clocks (circadian rhythmicity) and photoperiodic clocks (seasonal timing) (Helfrich-Förster 2002; Veerman 2001; Danks 2003, 2005; Bradshaw et al. 2006). Second, seasonal timing is thought to function by two separate measuring mechanisms. In studying seasonal timing in the spider mite, *Tetranychus urticae,* Veerman and VazNunes (1987) found that long days and short days are accumulated separately. Hardie & VazNunes (2001) found corresponding results in the aphids *Megoura viciae* and *Aphis fabae.* Kimura & Masaki (1998) studied the rather complicated diapause behavior of the cabbage moth, *Mamestra brassicae*, and found distinct differences between the short-day-induced summer diapause and the long-day-induced winter diapause. The authors postulated a regulation of the two responses by separate physiological processes.

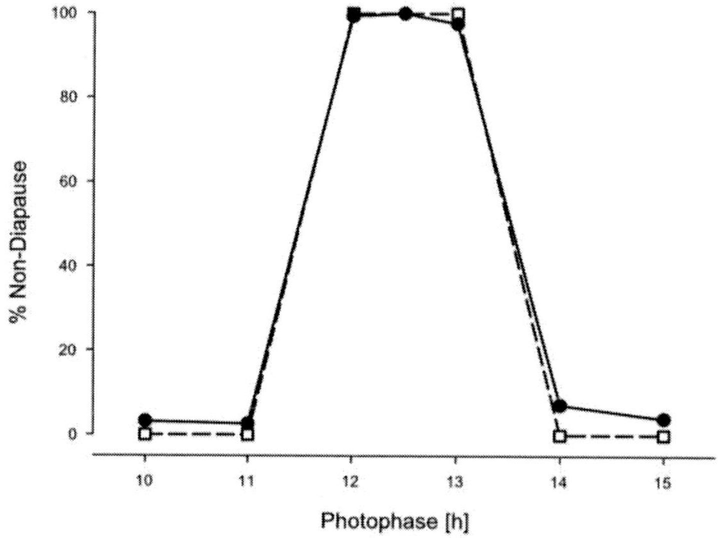

Figure 1. Photoperiodic response of *Pieris brassicae* in southwestern Europe at 18°C (solid line) and response modeled with the "double circadian oscillator clock" model (dashed line).

The diapause behavior of the large white butterfly, *Pieris brassicae*, is similar to that of *Mamestra brassicae*. Studies on the photoperiodic response of a population from southwestern Europe have led to rarely observed results (Figure 1). At 18°C, this species shows 100% winter diapause (hibernation) at short day lengths and 100% summer diapause (aestivation) at long day lengths. Both types of dormancy comply with a facultative diapause, meaning that each individual can react with aestivation, hibernation, or continuous development depending on the photoperiod. The observed variability in photoperiodic response results from an adaptation of the species to the specific climatic conditions in southwestern Europe (Spieth 2002). It has been proven beyond a doubt that *Pieris brassicae* measures diurnal light/dark changes quantitatively (Spieth & Sauer 1991). The absolute length of the scotophase determines the inductive strength of a given photoperiod (Veerman et al. 1988; Spieth 1995). Regarding quantitative measurement of the scotophase, it is incomprehensible how one measuring system only, working with the long day/short day principle, could lead to a diapause response under long-day as well as short-day conditions. As a result, one needs to ask whether aestivation and hibernation are controlled by the same measuring system or whether they are two independent reactions possibly controlled by two separate measurement systems.

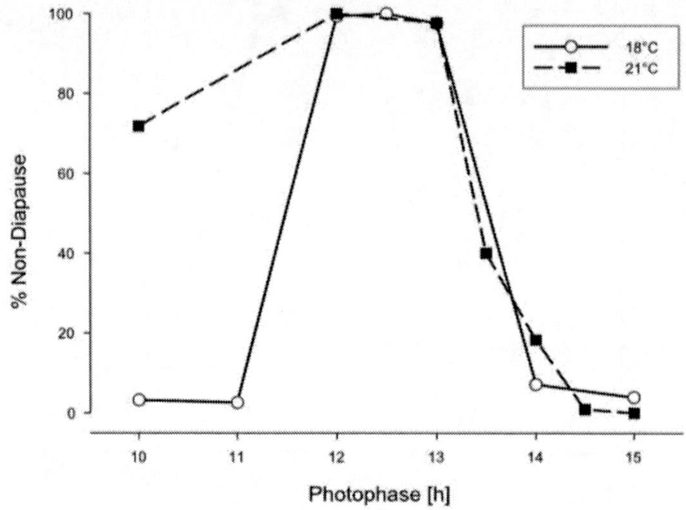

Figure 2. Temperature dependence of photoperiodic response of *Pieris brassicae* under hibernation vs. aestivation inducing photophases (from Spieth et al. 2004).

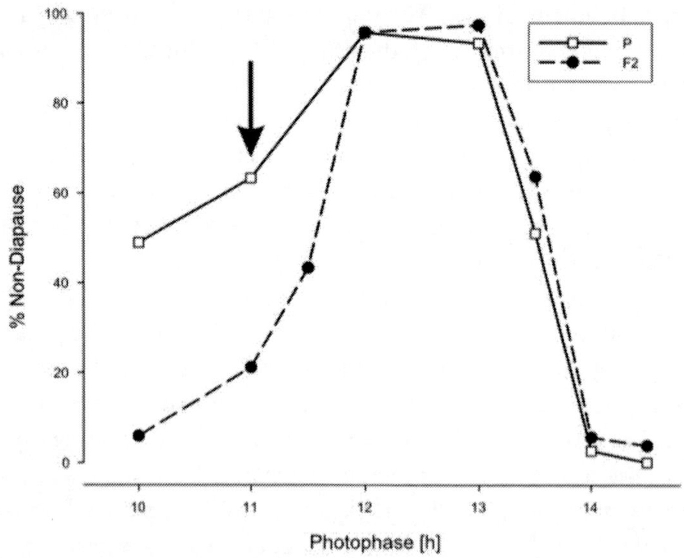

Figure 3. Change of photoperiodic response of *Pieris brassicae* after two inbreeding cycles (F2-generation) with individuals diapausing at 11h of light (arrow). Solid line indicates situation in the parental generation (P) (from Spieth et al. 2004).

Experimental Evidence for Two Unlinked Measurement Systems

If the two types of diapause induced by short days or long days are controlled by only one measuring system, a modification of one of the diapause responses is likely to influence the other diapause type. If, however, it can be shown that one diapause response can be influenced without any effect on the other, this would support the existence of two measurement systems, one controlling winter diapause, and the other, summer diapause.

Evidence supporting two systems comes from examination of the temperature sensitivity of the measuring system. In many species, the photoperiodic response is influenced by temperature (Danilevski 1965; Sauer et al. 1986). In experiments with *Pieris brassicae*, only hibernation was found to be highly temperature sensitive (Figure 2) whereas aestivation could not be influenced within the temperature range chosen in the experiment. In other species, too, the accumulation of short- and long-day events showed different levels of temperature dependence (Hardie 1990). The short-day response was found to be highly temperature compensated, contradicting the findings in *Pieris brassicae.* These results support the theory of two measurement systems in the photoperiodic timer.

The attribute for a comparison of photoperiodic responses between species, the so-called "critical photoperiod," at which 50% of all animals in a population show diapause, is a heritable trait. This has been shown for many species. By inbreeding certain response types, the critical photoperiod can easily be modified and shifted on the day-length axis (Sauer et al. 1986). This method was used to manipulate the hibernation response of *Pieris brassicae.*

If the measuring processes for aestivation and hibernation were coupled, a change of one dormancy response would lead to a change in the other, causing both to be altered. Individuals from a parental generation (P-generation in Figure 3) that had responded with a hibernation diapause at a photophase of 11 h were inbred. Photoperiodic response, as a heritable character, should therefore lead to an increase of the 11-h diapause type. After two inbreeding steps with the 11-h types of the P- and F1-generations, hibernation increased significantly (U-test: $p<0.001$). A comparison of the P- and F2-generation curves indicates a strong shift of the hibernation response to longer photophases, but no change in the aestivation response after two inbreeding cycles. This result does not provide evidence for a link between aestivation and hibernation responses.

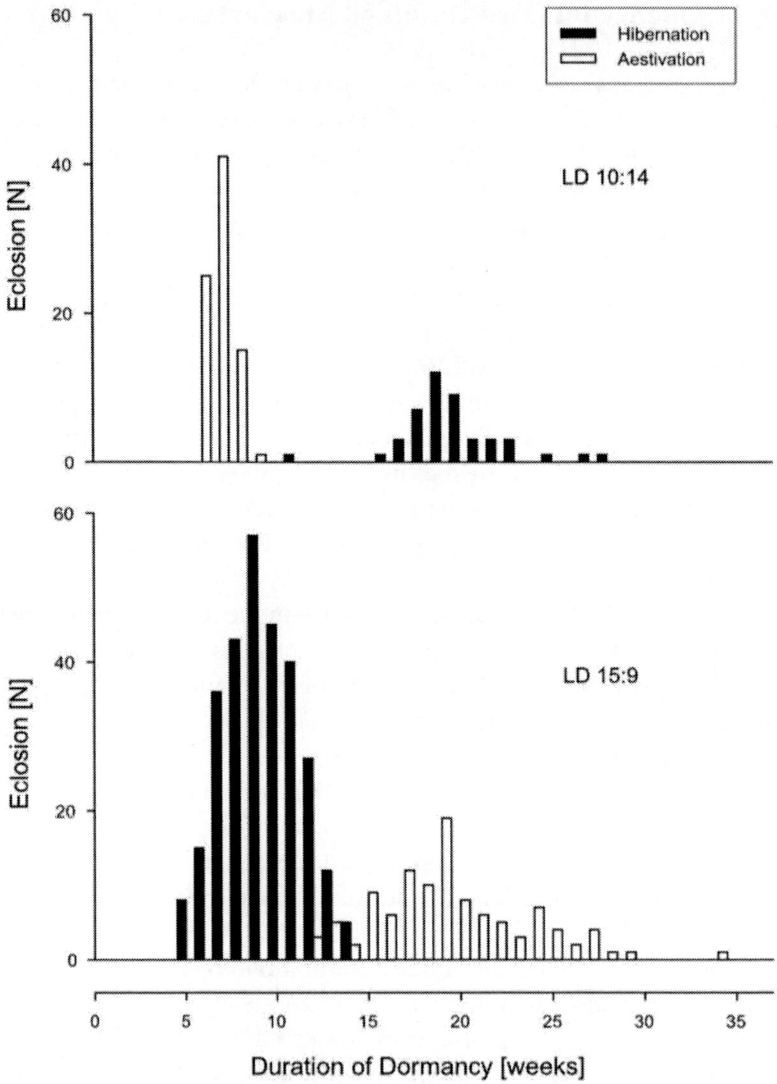

Figure 4. Duration of hibernation and aestivation diapause under long-day (LD 10:14) and short-day conditions (LD 15:9) at 21°C (from Spieth et al.2004).

More support for two separate measurement systems comes from the physiological state of diapausing individuals. If a diapause response, no matter whether hibernation or aestivation, is induced by the same mechanism, then it is very unlikely that diapausing individuals are in different physiological states (Kimura & Masaki 1998).

In *Pieris brassicae*, one subset of each dormancy type was continuously held in the dormancy-inducing photoperiodic condition after pupation, with 10 h of light in the hibernation type and 15 h in the aestivation type. A second subset of each type was crosswise transferred to the opposite rearing condition (Figure 4) and the time until butterfly eclosion was recorded. The state of summer dormancy was stabilized by long days. Eclosion started spontaneously after 12 weeks and peaked after 19 weeks. With a transition of the pupae to 10-h light conditions, aestivation synchronously terminated and butterfly eclosion occurred after 7 weeks. An almost inverse response was observed for winter dormancy. At 10 h of light, hibernation lasted for a mean of 19 weeks, and at 15 h, only 9 weeks. Hibernation and aestivation require opposite photoperiodic signals to terminate the respective dormancy stage. This implies that they are diapause types of different physiological quality. The most parsimonious explanation is thus to assume two independent measurement systems that put individuals into different physiological conditions.

Photosensitivity, "Required Day Number," and Perception of Increasing Day Length

Another option to search for differences between aestivation and hibernation responses is to examine the change of photosensitivity during ontogeny. For these experiments, the side conditions are very important and are therefore explained in detail now.

Usually, diapause can be induced only during very specific developmental stages (Danks 1987). In *Pieris brassicae* this is the larval stage only, as any of the other developmental stages are unsusceptible. To identify the photosensitive stage responsible for a particular reaction, a single change, from diapause-inducing to diapause-inhibiting photoperiods (or vice versa), is applied in most studies. The change is done in short intervals at different points of time during the ontogeny of the species studied, thereby allowing detection of the developmental phase that, in response, shows the largest effect. This phase is interpreted as that of highest photosensitivity. However, this frequently applied method only provides a broad estimate of photosensitivity, i.e., whether the sensitivity peak is more in the beginning, the middle, or at the end of a specific developmental phase. One of the reasons is that the inductive strength of a photoperiodic event depends on the sensitivity of the organism to the event; however, in many species, photosensitivity changes during development (Figure 5A) (Veerman 1977; Tauber et al. 1986; Kimura & Masaki 1998).

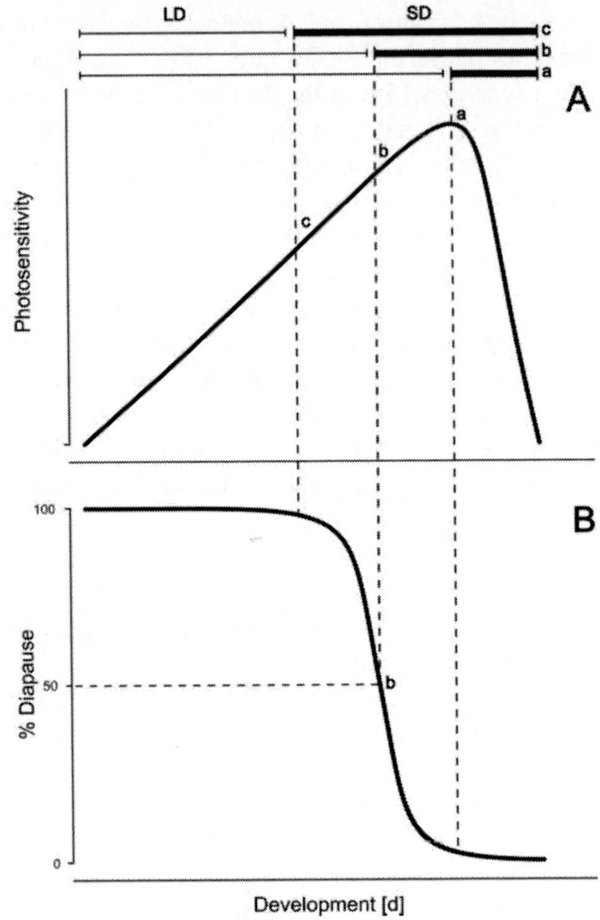

Figure 5. Schematic representation of the amount of diapausing individuals (B) by a 'one step change' from long day (LD) to short day (SD) at different times in development (top panel a, b, c). An early change (c) to short days results in diapause, a late (a) in non-diapause. The biggest change in diapause rate results from an LD/SD relation depicted as 'b'. This developmental stage is then interpreted as the one with highest sensitivity to photoperiods when the method of a single change in photoperiods is used. Curve A indicates the course of the continuously changing photosensitivity in e.g. *Pieris brassicae*, which points out that the situation 'b' is not the phase of highest photosensitivity. The false conclusion follows from the asymmetrical photosensitivity. Due to the longer ascending phase more longdays are experienced resulting in a major change of response before the sensitivity peak.

With changing photosensitivity it is impossible to predict a unique required day number (RDN) needed to induce diapause in different developmental phases. Therefore, to argue on the basis of RDN is very problematic.

With few exceptions (Adams 1985), the continuous change in photosensitivity during development was neglected in the past. Hypotheses invoking RDN, for example conclusions regarding the inductive strength of long days vs. short days (Hardie 1990) or the negligible influence of temperature on RDN (Saunders 1971), are not sufficiently supported by current knowledge. To propose such hypotheses it is necessary to consider parameters besides those commonly used to estimate RDN, as shown in the following.

The schematic drawing in Figure 5A assumes a continuously increasing photosensitivity that bottoms out at the end. This course approximates the photosensitivity of *Pieris brassicae* from the first to the fifth larval stage (Spieth 1995). Furthermore, it is assumed that several short days are needed to induce a diapause response. Due to the asymmetric course of the sensitivity curve, more photoperiodic events are needed to induce a response during phases of low sensitivity than during phases of high sensitivity. Not only does the number of short days decline if the change from long- to short-day conditions is shifted to a phase later in development (top panel in Figure 5A), but each additional long day has a stronger inductive value to prevent a diapause response (Figure 5A). A possible consequence of this is shown in Figure 5B. The change in photoperiodic response from diapause to non-diapause is clearly before the phase of highest photosensitivity. This is characterized by the 50% value. This value defines the point of "most effective change in diapause behavior," however, it does not correspond with the phase of highest photosensitivity. This means, if the experimental setup involves only one change in the sequence of photoperiods during development, misinterpretations of the course of photosensitivity can not be ruled out.

In this context, we also have to critically consider the putative ability of certain insect species to measure increasing and decreasing day lengths. By experiencing the same day lengths, increasing in spring, decreasing in fall, any individual can use these different signals to react according to the season. There are several studies apparently supporting this idea, at least according to the respective authors (Tauber et al. 1986; Saunders 2002; Nylin 1989). Yet all these studies have to be reassessed carefully. It is absolutely necessary not only to presume the course of photosensitivity indirectly (Tauber & Tauber 1970b) but also to know its course in every detail. Furthermore, it is crucial to investigate photosensitivity only within one developmental stage (egg, larvae, pupae, or imago) rather than work across all of them. If the experimental change in the photoperiod is stretched over two developmental stages (Tauber & Tauber 1970a, 1976), it has to be ensured that the sensitivity of both leads to the same reaction or behavior. The photosensitivity of pupae, for example, may be decisive for a completely different response than the one arising from the photosensitivity of larvae (Danks 1987; Kostal 2006). A photophase which induce diapause in larvae can have a contrasting effect in the pupal stage (see Figures 1 and 4), potentially leading to wrong conclusions when studies are extended across several ontogenetic stages.

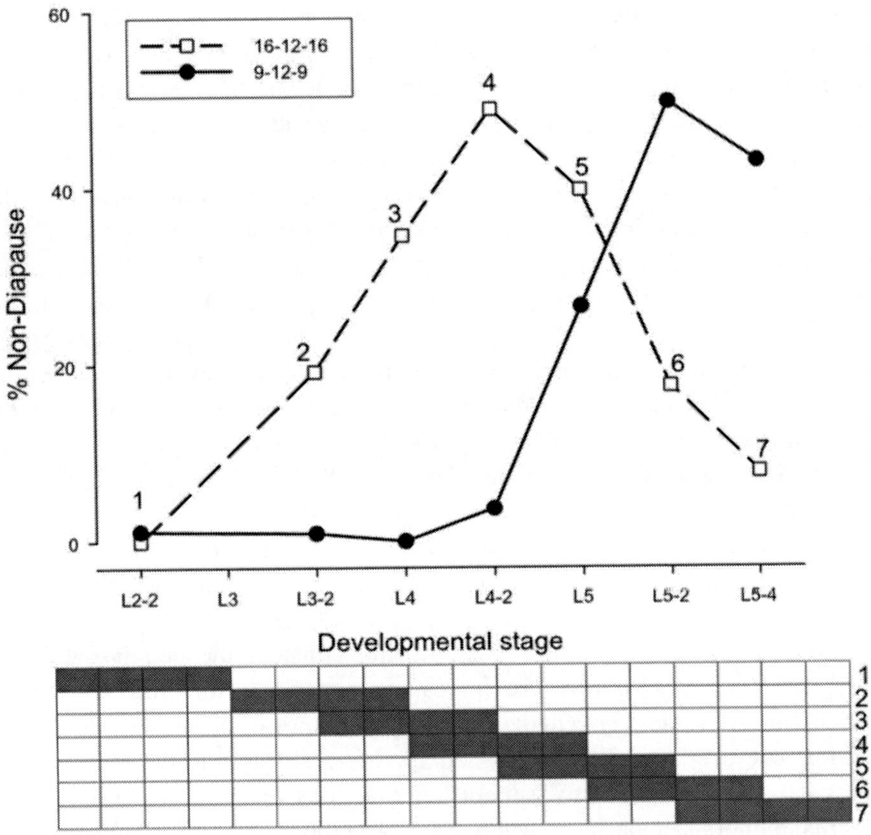

Figure 6. Change of photosensitivity of *Pieris brassicae* during larval development at 18°C, measured as non-diapause response. The background photoperiods were a hibernation inducing LD 9:15 (circles), or an aestivation inducing LD 16:8 (squares). Each background was scanned by 4 diapause-preventing events of intermediate LD 12:12 photoperiods (dark columns in schedule) (from Spieth et al. 2004).

Actual measurement of increasing or decreasing day lengths implies some type of comparative process, in terms of a photoperiod on day X in relation to that on day X+1. However, so far, we do not know of any physiological correlate for such a complex process. There is a much simpler explanation for the observed changes in photoperiodic response at increasing and decreasing day lengths. Usually, photosensitivity changes steadily within a certain developmental stage. The example of *Pieris brassicae* (Figure 5A) shows that the same photoperiod can have different efficiencies depending on its position (earlier or later) in the ascending phase of photosensitivity. Accordingly, if we consider a sequence of increasing day lengths, then, firstly, rising sensitivity will lead to the higher efficiency of every following day and, secondly, the steadily prolonging photophase itself will inhibit diapause more effectively from day to day. Conversely, if day length is decreasing during this developmental phase, the photophase at maximal photosensitivity will be shorter than that during the onset of photosensitivity and the ability to induce a non-diapause response will be reduced accordingly. Thus, if a species encounters the same sequence of photophases in the same developmental phase, in one case increasing and in the other decreasing, then the course of photosensitivity will inevitably lead to different photoperiodic responses. These are referred to in the literature under headings like "measurements of increasing and decreasing photophases." We can conclude that the ability to respond to naturally changing photoperiods does indeed exist and experimental evidence leaves no doubt about it. However, the measuring principles have nothing to do with the detection of increasing or decreasing day lengths; rather, they derive from the inconsistency of photosensitivity. It is a system-inherent mechanism without the need for any additional physiological process.

Photosensitivity as an Experimental Instrument for Investigating Clock Mechanisms

A single change in the sequence of a photoperiod is not sufficient to detect differences in sensitivity between hibernation and aestivation responses. To detect photosensitivity precisely, it is necessary to change the provided photoperiod twice. Ideally, a non-inductive photoperiodic background is interrupted by a single inductive event; that is, depending on the species, one short day or long day, and thereafter the individuals are brought back into the non-inductive conditions. Unfortunately, this is only possible in exceptional cases (Zaslavski & Kats 1988; Hardie 1990; Spieth 1995; Hua et al. 2005) because for the usual experimental setups more than one event is necessary to induce a response. Yet, as a rule of thumb, to estimate the phase of highest photosensitivity as precisely as possible, individuals should be exposed to the minimum number of inducing events.

This method was used to determine the course of photosensitivity in *Pieris brassicae* under hibernation and aestivation conditions. The photoperiodic response of this species at 18°C (Figure 1) indicates that photophases up to 11 h induce hibernation while those above 14 h induce aestivation. Photophases of 12–13 h inhibit diapause. For this experiment, photophases inducing either hibernation or aestivation were used as a photoperiodic background, which was then disrupted by four diapause-inhibiting photophases of 12 h light (see light schedule in Figure 6). The pattern of response was very similar for the two conditions; however, the period of strongest response clearly differed under hibernation and aestivation conditions. The most sensitive phase for aestivation was the fourth larval stage, and for hibernation, the fifth larval stage.

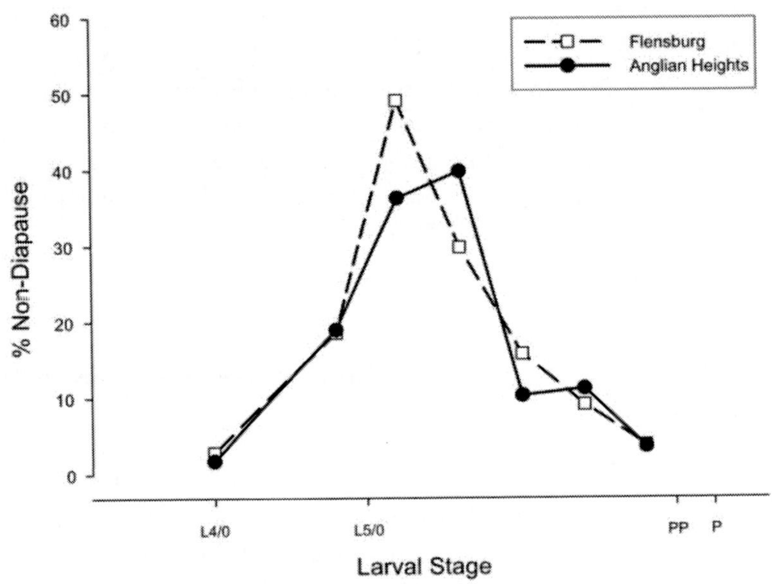

Figure 7. Comparison of the photoperiodic sensitivity in two non-aestivating populations of *Pieris brassicae* from Northern Germany (squares) and England (circles). Experimental setup as shown in Figure 6 with a higher temperature of 20°C. The diapause inducing background was interrupted by two long days of LD 19:6. In both populations sensitivity peaked one to two days after the ecdysis to L5 (modified from Spieth 1995).

Furthermore, marginal conditions have to be known to fully exclude any doubt that the observed difference is really based on the existence of two different measuring systems with different sensitivities. As shown in Figures 7 and 10, photosensitivity is a very stable species-specific trait without any geographic variation, and the phase of maximal photosensitivity is an inherent character valid for any photoperiod, no matter whether long-day or short-day (Figure 8). Additionally, there is no influence of temperature on the course of sensitivity (Figure 9). However, this conclusion is a somewhat intricate as the development time depends, of course, on the temperature, and the number of required days varies with temperature. In this sense, the experiments on *Pieris brassicae* have clearly established that the induction of a photoperiodic response does not depend on a RDN, but rather on whether inducing photoperiods are experienced by individuals during a certain developmental phase. It is irrelevant whether this developmental phase is prolonged by low temperature as long as it is illuminated by inducing photoperiods, no matter

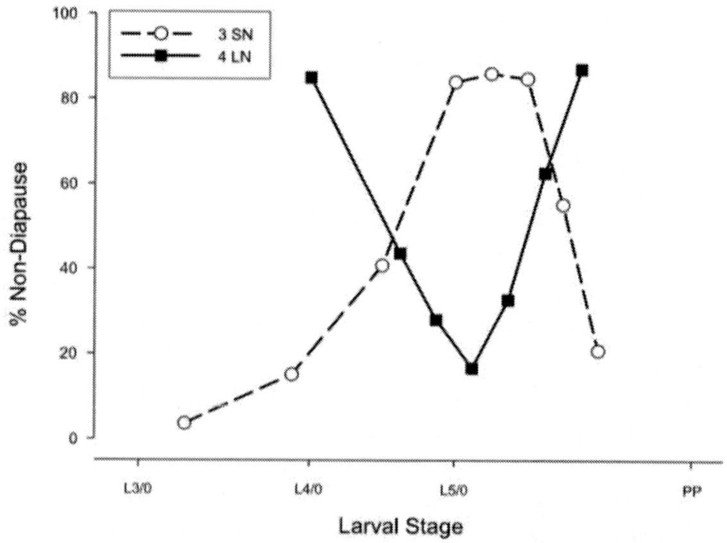

Figure 8. Comparison of photoperiodic sensitivity to long days (SN) and short days (LN) in a non-aestivating population of *Pieris brassicae* from Northern Germany. Sensitivity to long days was measured by interrupting a diapause inducing background with three long days of LD 19:5, sensitivity to short days by interrupting an non-diapause inducing background with 4 short days of LD 10:14. The general setup was as shown in Figure 6 (data from Spieth 1995).

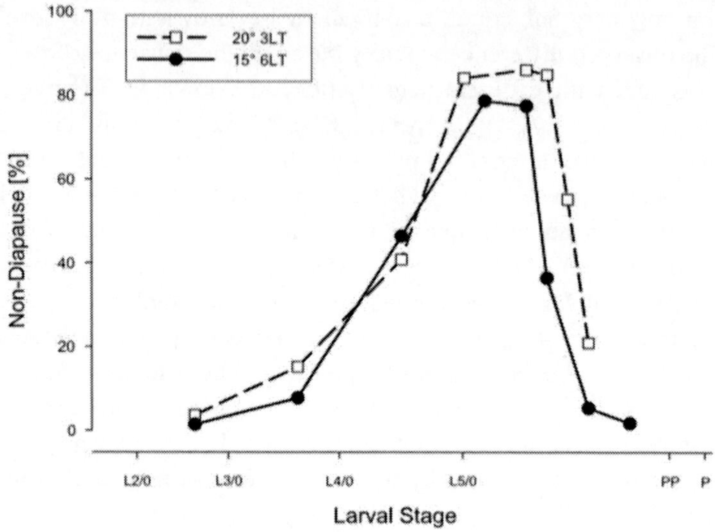

Figure 9. Influence of temperatures on photosensitivity in a non-aestivating population of *Pieris brassicae* from Northern Germany. The diapause inducing background was interrupted by three (20°C) and six (15°C) long days of LD 19:5. Sensitivity peaked at both temperatures at the first days after the ecdysis to L5. It lasts 5 days at 20°C from the L5/0 stage to the praepupa (PP) and 10 days at 15°C. The general setup was as shown in Figure 6 (data from Spieth 1995).

how long it might take (Spieth 1995). In *Pieris brassicae*, larval development at 15°C takes about twice as long as at 20°C, and the number of long days needed to induce a non-diapause response at 15°C is likewise twice the number needed at 20°C (Figure 9). Nonetheless, the only important fact is that the diapause-inhibiting photoperiod has to occur during the late 4th and early 5th larval stage. All these experiments indicate that photosensitivity is a very useful tool for investigating the photoperiodic measuring system.

These observations raise the question why, despite the numerous publications on RDN (Saunders 2002), it was not recognized that treatment during certain developmental stages, rather than the number of inductive events, is decisive for a photoperiodic response. One reason may be that more important than the difficulty in interpreting the RDN during different experimental treatments is the impact of the quantitative effect of a photoperiod, which is nearly always neglected and generally underestimated. The duration of a scotophase (or photophase) of a light/dark cycle can crucially influence the RDN. It is not without interest whether a short day has 12 or 13 h of darkness. To induce diapause in *Pieris brassicae*, for instance, scotophase of 13 h is more efficient than one of 12 h (Spieth & Sauer 1991). In the absence of a consideration of quantitative effects, statements about whether long days or short days are more efficient for a certain response are speculative (Hardie 1990; Hua et al. 2005). Several authors demonstrated that the efficiency of a photoperiod increases as the critical photophase is approached (Gibbs 1975; Kimura 1990; Hardie 1990; Kimura & Masaki 1992). An experiment in *Pieris brassicae* shows this in particular. The photosensitivity of a population from Denmark (Horsens) was investigated at 15°C (Figure 10). The background condition was a short day with a scotophase of 9 h 5 min, which in this Scandinavian population leads to 100% diapause induction. When the background was interrupted by six long-day events, the population hardly showed any reaction. When the scotophase of the shortday background was shortened by only 10 min, to a duration of 8 h 55 min, all individuals still showed diapause induction.

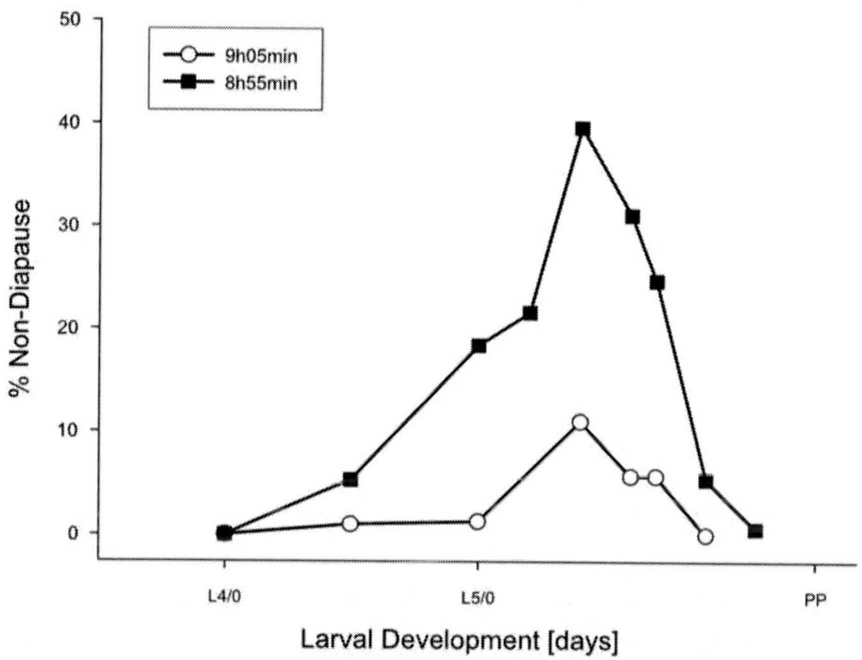

Figure 10. Importance of small changes in photoperiod for diapause response in *Pieris brassicae* measured with the experimental setup shown in Figure 6. Butterflies originated from a Danish population with a critical photoperiod of about LD 16:8. In the two experiments the photoperiodic background differed only by 10 minutes. The scotophase was 9h 5min and 8h 55min, respectively (see insert). See text for more details.

However, in this case the same sequence of six long days leads to a clearly interpretable reaction. Further experiments on *Pieris brassicae* have shown that even a change of only 5 min per day significantly alters the response (Spieth & Sauer 1991). These examples demonstrate the difficulty in using RDN as a dosage to interpret the photoperiodic responses if the quantitative effect of the photoregime is left unconsidered.

The issue becomes even more complex if the position of the photoperiodic response curve at different temperatures is considered. In many species, the curve is shifted along the day-length axis. If, for example, the temperature drops, then *Pieris brassicae* responds with non-diapause only at longer day lengths, that is at shorter scotophases. For a quantitative mechanism, this means that the same short day that was close to the critical photophases at higher temperatures, thus operating as a "long short day," is now further from the critical photophases at lower temperatures and its effect therefore changes to that of a "short short day." As noted before, to prove quantitative effects beyond any doubt, photoperiods close to the critical photophases (not to be mistaken for the critical photoperiod) have to be used, which in turn can lead to strange light/dark conditions, as seen in Figure 10.

Two Unlinked Measurement Systems

The differences between hibernation and aestivation diapause, summarized in Table 1, give ample support for the assumption that the responses are generated by different photoperiodic measurement systems.

Table 1. Differences in Aestivation vs. Hibernation response

Character	Effect	
	Hibernation	**Aestivation**
Temperature	strong	weak
Inbreeding response	strong	no effect
Duration at long-days	short	long
Duration at short-days	long	short
Peak Sensitivity	stage L5/2	stage L4/2
Inducing Photoperiod	short-days	long-days

Of all clock models published thus far (VazNunes and Saunders, 1999), only two are based on quantitative measurements and separate measuring systems. An early pioneer of a two-measurement hypothesis was Zaslavski (1987). Based on the principles of his more graphically modeled concept, VazNunes (1998) developed a mathematical model, which she named the "double circadian oscillator model." The fundamental assumption of this model is that there are two independent photoperiodic time measurement systems, one exclusively records short scotophases (or long days), the other, long scotophases (or short days). Both systems are based on an oscillating measurement principle. They are integrated into a single clock that processes the reciprocal signals of the two systems and calculates the relative dominance of each. The daily output of the clock determines the animal's final response.

These points can be summarized in a simple equation. The true signal that records the experience of a short night or long night is still unknown physiologically. As it involves cellular processes, it is assumed to be a not-yet-analyzed chemical substance. In the model, this is called substance C. The value of a photoperiod is defined by the amount of this substance, which in turn depends on the duration of the scotophase. For a short night, the value is called "shortval," for a long night it is "longval." Short nights and long nights are competing events that inhibit each other. The induction of diapause or non-diapause is, consequently, in accordance with the sum of all experienced photoperiodic events occurring during an animal's photosensitive phase. Thus, the induction sum, called INDSUM, is a product of the sum of all long-night values and all short-night values. It can be expressed as:

$$\text{INDSUM} = \Sigma\ \text{longval} - \Sigma\ \text{shortval} \qquad (1.1)$$

The mechanism for generating an oscillating measurement principle includes a time-delayed feedback loop as well as a system-inherent and a light-caused loss rate. Accordingly, both the amplitude and the oscillation period can be modified (VazNunes 1998). The simplified graphs in Figure 11 show an arbitrary oscillation with a period of 24 h. The model assumes that the short-night system only uses the ascending phase of the oscillation to produce the short-night substance CSN whereas the long-night system produces the equivalent long-night substance CLN only during the descending phase of the oscillation. Figure 11A relates to the short-night measurement system. Only if CSN exceeds a certain threshold (thresh) the event is accumulated. The maximal value for CSN (CSNmax) is reached if no darkness falls on the ascending phase of the oscillation. If the scotophase of a short night is very short, then CSN is only slightly reduced (example 1 in Figure 11A). The longer the scotophase, the less substance is synthesized. If the scotophase reaches or exceeds CSNmax then this day will not allocate any short-night

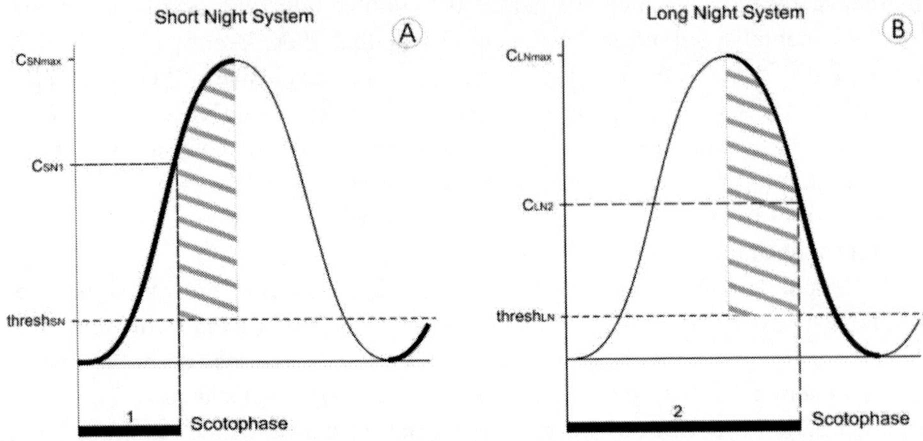

Figure 11. Schematic basics of the two measurement systems used in the "double circadian oscillator clock". The shaded areas indicate the amount of substance synthesized in case 1 by the short night system and case 2 by the long night system. For further details see text.

substance to the "counter" of the clock. The short-night process can generally be described as:

$$\text{shortval: } C_{SN} = (C_{SNmax} - C_{SN}(t)) \times SNV \quad (1.2)$$

where SNV is a constant derived from modeling the oscillation (VazNunes 1998).

The long-night system starts producing the long-night substance with the onset of the descending oscillation (Figure 11B). If the scotophase does not exceed the peak of the oscillation, no CLN substance is produced. The maximal value for CLN is reached if the scotophase lasts up to the intersection of the descending oscillation and the threshold. This process can be described with:

$$\text{longval: } C_{LN} = (C_{LNmax} - C_{LN}(t)) \times LNV \quad (1.3)$$

where LNV is the long-night equivalent of SNV in Equation 1.2.

When we put the exact terms into Equation 1.1, the induction sum is:

$$INDSUM = \Sigma\, (C_{LNmax} - C_{LN}(t)) \times LNV - \Sigma\, (C_{SNmax} - C_{SN}(t)) \times SNV \quad (1.4)$$

In other words, the longer the scotophase of a so-called long day the weaker its inductive strength, meaning the smaller the amount of short-night substance produced. Likewise, the longer the scotophase of a so-called short day the stronger its inductive strength as the amount of long-night substance rises. This principle only slightly deviates from the original model by VazNunes (1998) in the estimate of shortval.

The characteristics of the photoperiodic clock can be summarized as follows: the clock comprises two measurement systems, a long-night system and a short-night one. The two systems are unlinked. In nature, the daily output of both systems changes with natural day length. The daily output is integrated in the induction sum. The resulting photoperiodic response depends on the dominance of either longval or shortval. If one system dominates over the other and exceeds the induction threshold, a diapause response is initiated.

Evidence for the Model

Certainly, one can question the relevance of the model for natural systems. It has to be said that all models for the photoperiodic clock are based on mathematical descriptions of the photoperiodic response of a certain species. If a model also explains the photoperiodic responses of other species, its general validity is assumed (see Figure 1). Thus, the modeled photoperiodic response is verified by photoperiodic responses. Surely the relevance of a model is greater the more cases that can be explained by it, but ultimately these procedure do not prove the model.

Animals from an Iberian population of *Pieris brassicae* made it possible to test certain predictions of the "double circadian oscillator model" by a method other than the usual photoperiodic response curves, i.e., by determination of the photosensitivity. If we link the photoperiodic response of the Iberian population (Figure 1) to the measurement systems described, the response can be divided into two measuring processes. Figure 12 shows the characteristics of how the two measuring systems work schematically. The hibernation response is dominated by the long-night-determining system while the aestivation response is dominated by the short-night-determining system. If this assumption holds true, the long-night system dominates at scotophases longer than 13 h and leads to 100% hibernation. Between 12 h and 11 h, the two systems neutralize each other resulting in 100% non-diapause. The short-night systems dominate at scotophases shorter than 10 h and lead to 100% aestivation.

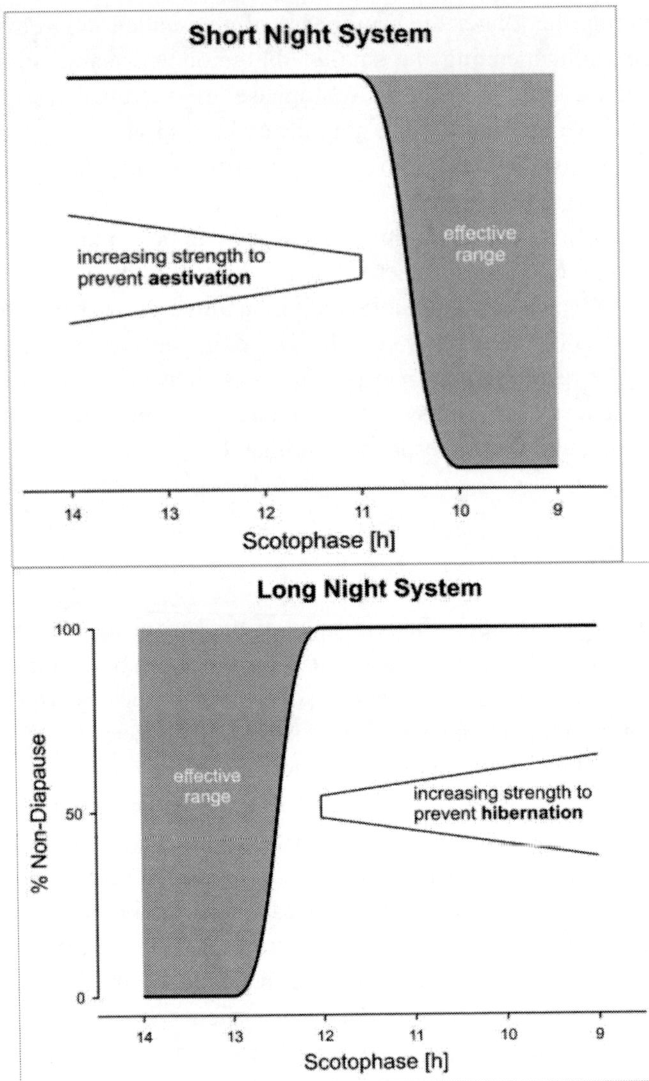

Figure 12. Schematic diagram of daylength dependent characteristic in both measurement systems. The effective range of scotophases inducing a hibernation diapause by the long night measurement system is marked by the shaded area in the left panel. Scotophases shorter than 12h have an increasing strength to prevent hibernation in *Pieris brassicae*. The effective range of scotophases inducing an aestivation diapause by the short night system is marked by the shaded area in the right panel. Scotophases longer than 11h have an increasing strength to prevent aestivation.

It is impossible to test these assumptions with the classical methods used to study photoperiodic responses, but they can be verified by investigating photosensitivity. To find differences between the photosensitivity of hibernation and aestivation (Figure 6) the diapause-inducing background (L/D 9/15 h or L/D 16/8 h) was interrupted with a four-day sequence of non-diapause-inducing intermediate photoperiods (L/D 12/12 h). From Figure 12 it can be deduced that, in case of the hibernation-inducing long-night measurement system, not only intermediate scotophases of 12 h but any scotophase below 12 h, even those inducing an aestivation response, should prevent hibernation diapause.

This assumption was tested using the same setup as in Figure 6. This time, the long-night background (L/D 9/15 h) was interrupted with four short nights of L/D 16/8 h and not, as in Figure 6, with intermediate photoperiods of L/D 12/12 h. If the photoperiodic response is controlled by a single measurement system, then no diapause-free development should be detected as both photoregimes, L/D 9/15 and L/D 16/8, induce diapause under natural conditions (Figure 1). However, the result (Figure 13) supports the two-measurement system. The photosensitivity for the hibernation response can be detected not only with diapause inhibiting photoperiods but also with aestivation inducing photoperiods.

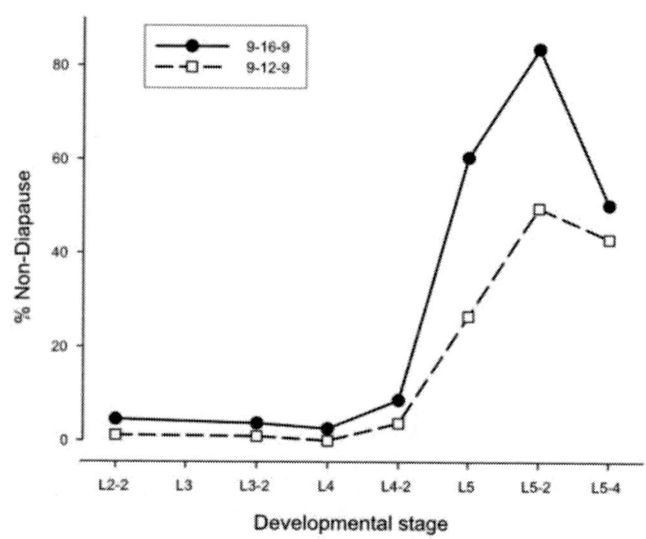

Figure 13. Change of larval photosensitivity tested on basis of the experimental setup shown in Figure 6. The hibernation inducing background (LD 9:15) was interrupted by 4 LD 16:8 cycles (solid line). The non-diapause effect of this normally aestivation-inducing photoperiod is compared with the response to 4 diapause-preventing LD 12:12 cycles (dashed line) (from Spieth et al. 2004).

Testing the short-night measurement system in the same way leads to a similar conclusion. If the summer diapause-inducing background (L/D 16/8 h) is interrupted not with four intermediate photoperiods (L/D 12/12 h) but with four L/D 9/15 h, very similar non-diapause responses can be observed (Figure 14). Here as well we found that the photosensitivity of the short-night measurement system can be examined successfully by using photoperiods that usually lead to a long-night diapause. This means that the same photoperiod is able to induce diapause or non-diapause, as the photoperiodic context in which an animal experiences the event is important.

In both photosensitivity tests, the response is stronger than if induced by intermediate photoperiods (dashed curves in Figures 13. and 14). This means that, compared to the general photoperiodic response (Figure 1), photoperiods that normally induce diapause 1) inhibit diapause in the photosensitivity test and 2) do this more efficiently than the naturally diapause-inhibiting light/dark cycles (L/D 12/12; 11/13). The "double circadian oscillator model"

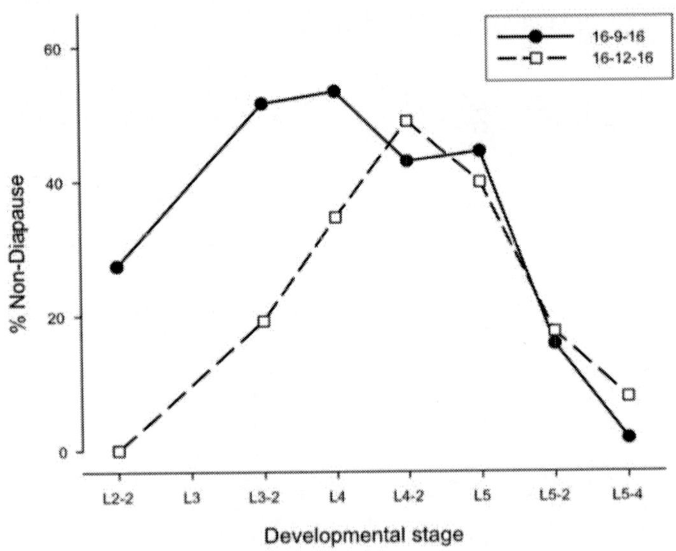

Figure 14. Change of larval photosensitivity tested on basis of experimental setup shown in Figure 6. The aestivation inducing background (LD 16:8) was interrupted by 4 LD 9:15 cycles (solid line). The non-diapause effect of this normally hibernation-inducing photoperiod is compared with the response to 4 diapause-preventing LD 12:12 cycles (dashed line) (from Spieth et al. 2004).

(Equation 1.4; Figure 11) provides a plausible explanation. The long-night measuring system synthesizes a specific quantum of CLN at L/D 9/15 h. If interrupted with four short nights (L/D 16/8 h or L/D 12/12 h), the amount of CLN at these days is zero. These four events diminish the total amount of CLN at the end of larval development. If the short nights are applied during the phase of high photosensitivity (fifth larval stage, L5) the loss of CLN is particularly high and the threshold value of CLN for a hibernation response is only reached in some individuals (non-diapause peak in Figure 13). The model further predicts that during short nights not only is the synthesis of CLN zero but the short-night-determining system produces the antagonistic CSN, which additionally diminishes the amount of accumulated CLN. Compared to L/D 12/12 h, the shorter scotophase of the L/D 16/8 h "short night" is a stronger event, producing a higher amount of CSN, and therefore has a stronger avoiding effect on hibernation. This results in a larger number of non-diapause individuals (Figure 13). Under reverse photoperiodic conditions (Figure 14), the same applies for processes inducing aestivation.

Summing up, we can state that the findings for the photoperiodic response of *Pieris brassicae* support the "double circadian oscillator" model. Two independent time measurement systems, one determining long nights (LN) (short-day sensitive) and the other short nights (SN) (long-day sensitive), are involved. Within its effective range, each system synthesizes a hypothetical substance (CLN by the LN system; CSN by the SN system). The daily amount of substance generated by each system varies depending on the length of the scotophase. In addition, the daily produced substance accumulates during the photosensitive larval stages. The effective range of the LN-measuring system is limited to long nights, that of the SN-measuring system to short nights. Beyond its range of effectiveness the output of a system is zero. Moreover, the two systems act as antagonistic processes. The photoperiodic response of an individual is threshold-dependent. The model describes quite different features for the two measuring systems but they render explainable all of the above-mentioned phenomena. The increase in hibernation by the inbreeding of "long-night response types" (Figure 3) selects the LN-measuring system while leaving the SN system untouched, with the consequence that the hibernation response shifts on the day-length axis without affecting the aestivation response. The evaluation of the temperature effect yields a similar result. The SN system is largely temperature-compensated, which is reflected by a quite steady aestivation response, in contrast to the high temperature dependence of the LN system, with its correspondingly variable hibernation response (Figure 2). Finally, the LN system has its highest photosensitivity in the L5 stage, the SN system in the L4 stage (Figure 6). If CLN and CSN are different substances, this will contribute to explaining the different physiological states of aestivating and hibernating pupae (Figure 4).

Perspective

Geographic variation of the photoperiodic response

The photoperiodic response of *Pieris brassicae*, with winter and summer diapauses (Figure 1), is only found in southwestern Europe (Held & Spieth 1999; Spieth & Schwarzer 2001). In northern Spain, this species shows no summer diapause but instead a continuous development under short-night conditions. This poses the question whether the dramatic change in photoperiodic response requires major modifications in the photoperiodic time measurement (Spieth et al. 2004). The "double circadian oscillator" model provides an explanation with only minor modifications. We illustrate the mechanism by a more visual approach using Figures 11 and 12. The threshold determining the maximally synthesized amount of substance C is variable in each of the two measurement systems. Accordingly, a rise in the threshold for the SN-determining system would lead to a smaller effective amount of synthesized CSN each day. This, in turn, would have the effect that at the end of the sensitive phase the induction sum for the summer diapause is not reached and the animals develop without diapause. This is a very simple explanation, derived from the model, for a photoperiodic response without a summer diapause.

Continuing along these lines, we come to the conclusion that the heritable parameter of the photoperiodic response is not the critical photoperiod (Beck 1980; Sauer et al. 1986; VazNunes & Hardie 1993). Physiologically controlled switching events, like diapause vs. non-diapause, are usually threshold-dependent. A variation in the threshold is the basis for selective processes leading to the adaptation of populations to changing ecological conditions. In the case of geographic variations, such processes are frequently expressed in the photoperiodic responses of different species (e.g., Danilevski 1965; Tauber et al. 1986; Zaslavski 1988). In our study, these responses comprised the thresholds of the SN- and LN-determining systems as well as the induction sum. If we exclude the length of scotophase as an extrinsic factor, the thresholds of the two measurement systems are the crucial parameters that define the dominance of the systems. Finally, the threshold for the induction sum decides between diapause and non-diapause. As a result, the variance in the photoperiodic response of individuals in a population is then an expression of the variance in these thresholds. In the pitcher-plant mosquito, *Wyeomyia smithii*, Wegis et al. (1997) interpreted the clinal shift of the photoperiodic response curve by a change in the amplitude of the response. This is consistent with the assumption of a geographically continuous changing induction sum.

The photoperiodic response curve of a species is the image of the induction sum at the end of the photosensitive phase.

Is Pieris brassicae an exception or a model for a general principle?

If we pose this question and search for experimental evidence provided by the "double circadian oscillator" model, we soon arrive at the methodological limits. *Pieris brassicae* is an especially lucky case, as it shows a day-length-controlled, 100% facultative diapause under long- and short-day conditions (Figure 1). There are only a few other species known to respond in a similarly clear way. But this unambiguous behavior in response to long and short days is a precondition for reasoning based on the photosensitivity test. Only then can the antagonistic effects of these photoregimes on one measurement system be tested with absolute confidence. The photosensitivity test has not yet been applied to any other species. A very suitable test organism, for example, would be the cabbage moth, *Mamestra brassicae*, a species widespread in Japan (Masaki 1968) and Europe (Grüner & Sauer 1988). The moth exclusively hibernates in its northern distribution ranges and exhibits 100% aestivation in its southern ranges. In the studies published in the 1990s, not all of the moth's photoperiodic responses could be explained by the existing models (Kimura & Masaki 1992, 1998). The behavior of the cabbage butterfly, *Pieris melete*, found in China, is similar to that of *Pieris brassicae* (Xue et al. 1997; Xiao et al. 2008), making it very suitable for further tests of the "double circadian oscillator" model.

When the question whether *Pieris brassicae* is an exception or not is answered by using photoperiodic response curves, the answer is clearly no. Danks (1987), for example, listed several species exhibiting dormancy under long- and short-day conditions. Beck (1980) characterized this as a "type III" reaction. However, this and similar reactions can belong to other forms of developmental arrest (Beck 1980; Kostal 2006), as observed in *Pieris brassicae*, and such dormancies are often exhibited under unnatural day lengths. But these reactions show clearly that dormancies under short- and long-day conditions in one species are not rare.

Moreover, there are many species that only exhibit winter or summer diapause (Masaki 1980; Zaslavski 1988). In these cases we have to consider that geographically caused variations in the species were often not tested, and the experimental conditions necessary to reveal both dormancy types neglected. In addition to day length, temperature has an important influence on the characteristics of the photoperiodic response (Tauber et al. 1986; Danks 1987). Aestivation and hibernation often occur only within certain temperature ranges (Xue et al. 1997; Held & Spieth 1999; Xiao et al. 2008), and it is not unusual that the ability of a species to exhibit both dormancy types is revealed only under ecologically extreme conditions. In the cotton bollworm, *Helicoverpa armigera*, a summer diapause cannot be induced until temperatures climb above 33°C (Liu et al. 2006). Even at these high temperatures, this species only exhibits a partial summer diapause, which means that only some individuals show an aestivation response. This mixed reaction is common for species exhibiting both dormancy types (Grüner & Masaki 1994; Wu et al. 2006; Xiao et al. 2008).

We do not mean to give species-specific details but rather seek to highlight the fact that many species are able to respond to short and long days with developmental arrest. Clinal changes in the photoperiodic response, from typical summer diapause in southern regions to absolutely no response in northern regions, are important case studies. They document that even northern populations of a species without a detectable aestivation behavior must have a cryptic potency for a summer diapause response under long-day conditions (Imai 2004). The geographic transition zone from non-aestivating to aestivating populations can be very broad, as seen in *Mamestra brassicae* (Grüner & Masaki 1994), or, as in *Pieris brassicae*, it may not be any wider than 200 km (Spieth & Schwarzer 2001). There are also very special cases, such as *Chlorops oryzae*, which exhibits a bivoltine life cycle due to a summer diapause but in some areas, due to suppressed aestivation, a trivoltine cycle (Takeda 1997). Such organisms are examples of the intraspecific variance in which both dormancy types are exhibited.

All these facts are reason enough to conclude that the photoperiodic time measurement system in arthropods could generally be based on two measuring systems. Further experimental support for this hypothesis by photosensitivity tests is possible with only a few other species. In species in which only one dormancy response can be induced, it will be impossible to test experimentally whether the species is fundamentally unable to exhibit both diapause responses or whether the dominance of one system completely suppresses the other. The best hope of clarifying this issue is to unravel the physiological processes of photoperiodic timing.

Our aim in this chapter has been to emphasize the need to consider two independent photoperiodic time measurement systems when interpreting genetic or molecular findings. We think this is the best way to avoid inconsistencies and contradictions.

REFERENCES

Adams, J. A. (1985). The photoperiodic induction of ovarian diapause in the cabbage whitefly, *Aleyrodes proletella* (Homoptera: Aleyrodidae). *Journal of Insect Physiology, 31*, 693-700.

Beck, S. D (1980). *Insect Photoperiodism.* 2nd edn. Academic Press, New York.

Bradshaw, W. E. Quebodeaux, M. C., Holzapfel, C. M. & Mathias, D. (2003).. The condribution of an hourglass timer to the evolution of photoperiodc response in the pitcher-plant mosquito, *Wyeomyia smithii. Evolution, 57 (10)*, 2342-2349.

Bradshaw, W. E., Holzapfel, C. M. & Mathias, D. (2006). Circadian rhythmisity and photoperiodism in the pitcher-plant mosquito: can the seasonal timer evolve independently of the circadian clock? *American Naturalist., 167(4)*, 161-165.

Bünning, E. (1958). *Die physiologische Uhr.* Springer, Berlin.

Danilevskii, A. S. (1965). *Photoperiodism and Seasonal Development of Insects.* Oliver & Boyd, London.

Danks, H. V. (1987). *Insect Dormancy: An ecological perspective.* Biological Survey of Canada, Ottawa.

Danks, H. V. (2003). Studying insect photoperiodism and rhythmicity: components, approaches and lessons. *European Journal of Entomology 100*, 209-221.

Danks, H. V. (2005). How similar are daily and seasonal biological clocks? *Journal of Insect Physiology, 51*, 609-619.

Dumortier, B. & Brunnarius, J. (1987). Photoperiodic time measurement in an insect: is the photoinducible phase φ_i a reality? In: Pévet ed.: *Circadian Rhythms, Photoperiod and Endocrine Secretion.* Comparative Physiology of Environmental Adaptations, vol.*3*. (8th ESCP Conference, Strasbourg 1986). Karger, Basel; pp. 1-22.

Dumortier, B. (1994). The "Circadian Paradigma": a test of involvement of the circadian system in the photoperiodic clock. *Journal of Theoretical Biology, 166*, 101-112.

Gibbs, D. (1975). Reversal of pupal diapause in *Sarcophora argyrostoma* by temperature shifts after puparium formation. *Journal of Insect Physiology 21*, 1179-1186.

Grüner, C. & Sauer, K. P. (1988). Aestival dormancy in the cabbage moth *Mamestra brassicae* L. (Lepidoptera: Noctuidae). 1. Adaptive significance of variability of two traits: Day length thresholds triggering aestival dormancy and duration of aestival dormancy. *Oecologia, 74,* 515-523.

Grüner, C. & Masaki, S. (1994). Summer diapause in the polymorphic life cycle of the noctuid moth *Mamestra brassicae*. In: Danks, HV ed.: *Insect Life-Cyle Polymorphism. Theory, evolution, and ecological consequences for seasonality and diapause control*. Kluwer Academic Publishers, Dordrecht; pp., 191-204.

Hardie, J. (1981). Juvenile hormone and photoperiodically controlled polymorphism in *Aphis fabae*: postnatal effects on presumptive gynaparae. *Journal of Insect Physiology, 27*, 347-355.

Hardie, J. (1990). The photoperiodic counter, quantitative daylength effects and scotophase timing in the vetch aphid *Megoura viciae*. *Journal of Insect Physiology, 27*, 773-777.

Hardie, J. & Vaz Nunes, M. (2001). Aphid photoperiodic clocks. *Journal of Insect Physiology, 47*, 821-832.

Held, C. & Spieth, H. R. (1999). First evidence of pupal summer diapause in *Pieris brassicae* L.: the evolution of local adaptedness. *Journal of Insect Physiology, 45*, 587-598.

Helfrich-Förster, C. (2002). The circadian system of *Drosophila melanogaster* and its light input pathways. *Zoology, 105*, 297-312.

Hua, A., Yang, D., Wu, S. & Xue, F. S. (2005). Photoperiodic control of diapause in *Pseudopidorus fasciata* (Lepidoptera: Zygaenidae) based on qualitative time measurment. *Journal of Insect Physiology, 51*, 1261-1267.

Hua, A., Xue, F. S., Xiao, H. J. & Zhu, X. F. (2005). Photoperiodic counter of diapause induction in *Pseudopidorus fasciata* (Lepidoptera: Zygaenidae). *Journal of Insect Physiology, 51*, 1287-1294.

Imai, C. (2004). Photoperiodic induction and termination of summer diapause in adult *Epilachna admirabilis* (Coleoptera: Coccinellidae) from a warm temperate region. *European Journal of Entomology, 101*, 523-529

Kimura, M. T. (1990). Quantitative response to photoperiod during reproductive diapause in the *Drosophila auraria* species-complex. *Journal of Insect Physiology, 36*, 147-152.

Kimura, Y. & Masaki, S. (1992). Effect of light period on dark-time measurement for diapause induction in *Mamestra brassicae*. *Journal of Insect Physiology, 38*, 681-686.

Kimura, Y. & Masaki, S. (1998). Diapause programming with variable critical daylength under changing photoperiodic conditions in *Mamestra brassicae*

(Lepidoptera: Noctuidae).. *Entomological Science, 1*(*4*), 467-475.

Koštál, V. (2006). Eco-physiological phases of insect diapause. *Journal of Insect Physiology, 52*, 113-127.

Koštál, V. & Hodek, I. (1997). Photoperiodism and control of summer diapause in the mediterranean tiger moth *Cymbalophora pudica*. *Journal of Insect Physiology, 43*, 767-777.

Koštál, V., Shimada, K. & Hayakawa, Y. (2000). Induction and development of winter larval diapause in a drosophilid fly, *Chymomyza costata*. *Journal of Insect Physiology, 46*, 417-428.

Kroon, A., Veenendaal, R. L. & Veerman, A. (1997). Photoperiodic induction of diapause in the spider mite *Tetranychus urticae*: qualitative or quantitative time measurement? *Physiological Entomology, 22*, 357-364.

Lees, A. D. (1955). *The Physiology of Diapause in Arthropods*. Cambridge University Press, Cambridge.

Lewis, R. D. & Saunders, D. S. (1987). A damped circadian oscillator model of an insect photoperiodic clock. I. Description of the model based on a feedback control system. *Journal of Theoretical Biology, 128*, 47-59.

Liu, Z., Gong, P., Wu, K., Sun, J. & Li, D. (2006). A true summer diapause induced by high temperatures in the cotton bollworm, *Helicoverpa armigera* (Lepidoptera: Noctuidae). *Journal of Insect Physiology, 52*, 1012-1020.

Masaki, S. (1968). Geographic adaptation in the seasonal life cycle of *Mamestra brassicae* (Linné) (Lepidoptera: Noctuidae). *Bulletin of the Faculty of Agriculture Hirosaki University, 14*, 16-26.

Masaki, S. (1980). Summer diapause. *Annual Review of Entomology* 25, 1-25.

Nylin, S. (1989). Effects of changing photoperiods in the life cycle regulation of the comma butterfly, *Polygonia c-album* (Nymphalidae). *Ecological Entomology, 14*, 209-218.

Sauer, K. P., Spieth, H. R. & Grüner, C. (1986). Adaptive significance of genetic variability of photoperiodism in Mecoptera and Lepidoptera. In: Fritz Taylor & Richard Karban editors. *The evolution of insect life cycles*. Springer-Verlag, New York, Berlin; pp. 151-172.

Saunders, D. S. (1971). The temperature-compensated photoperiodic clock "programming" development and pupal diapause in the flesh-fly, *Sarcophaga argyrostoma*. *Journal of Insect Physiology, 17*, 801-812.

Saunders, D. S. (2002). *Insect clocks*. 3rd ed. Elsevier Science, Amsterdam, 576pp.

Saunders, D. S. (2005). Erwin Bünning and Tony Lees, two giants of chronobiology, and the problem of time measurement in insect photoperiodism. *Journal of Insect Physiology, 51*, 599-608.

Saunders, D. S. & Lewis, R. D. (1987). A damped circadian oscillator model of an insect photoperiodic clock. II. Simulations of the shape of the photoperiodic response curves. *Journal of Theoretical Biology, 128*, 61-71.

Saunders, D. S., Lewis, R. D. & Warman, G. R. (2004). Photoperiodic induction of diapause: opening the black box. *Physiological Entomology, 29*, 1-15.

Spieth, H. R. (1995). Change in photoperiodic sensitivity during larval development of *Pieris brassicae* . *Journal of Insect Physiology, 41*, 77-83.

Spieth, H. R. (2002). Estivation and hibernation of *Pieris brassicae* (L.) in southern Spain: synchronization of two complex behavioral patterns. *Population Ecology, 44*, 273-280.

Spieth, H. R. & Sauer, K. P. (1991). Quantitative measurement of photoperiods and its significance for the induction of diapause in *Pieris brassicae* (Lepidoptera, Pieridae). *Journal of Insect Physiology, 37*, 131-138.

Spieth, H. R. & Schwarzer, E. (2001). Aestivation of *Pieris brassicae* (Lepidoptera: Pieridae): Implications for parasitism. *European Journal of Entomology, 98*, 171-176.

Spieth, H. R., Xue, F. S. & Strauß, K. (2004). Induction and inhibition of diapause by the same photoperiod: Experimental evidence for a "double circadian oscillator clock". *Journal of Biological Rhythms, 19*, 483-492.

Tauber, M. J. & Tauber, C. A. (1970a). Photoperiodic induction and termination in an insect: response to changing day lengths. *Science,* 167, 170.

Tauber, M. J. & Tauber, C. A. (1970b). Adult diapause in *Chrysopa carnea*: stages sensitive to photoperiodic induction. *Journal of Insect Physiology, 16*, 2075-2080.

Tauber, M. J. & Tauber, C. A. (1976). Developmental requirements of the univoltine species *Chrysopa downesi*: photoperiodic stimuli and sensitive stages. *Journal of Insect Physiology, 22*, 331-335.

Tauber, M. J., Tauber, C. A. & Masaki, S. (1986). *Seasonal Adaptations of Insects.* Oxford University Press, New Yorck.

Vaz Nunes, M. (1998). A double oscillator model for quantitative photoperiodic time measurement in insects and mites. *Journal of Theoretical Biology, 194*, 299-311.

Vaz Nunes, M. & Hardie, J. (1993). A model for the photoperiodc clock in the aphid *Megoura viciae*. *Journal of Insect Physiology, 39*, 173-182.

Vaz Nunes, M. & Saunders, D. S. (1999). Photoperiodic time measurement in insects: A review of clock models. *Journal of Biological Rhythms, 14*, 84-104.

Veerman, A. (1977). Aspect of the induction of diapause in a laboratory strain of the mite *Tetranychus urticae*. *Journal of Insect Physiology, 23*, 703-711.

Veerman, A. (1994). Photoperiodic and termoperiodic control of diapause in plant-inhabiting mites: a review. *Netherlands Journal of Zoology, 44*, 139-155.

Veerman, A. (2001). Photoperiodic time measurement in insects and mites: a critical evaluation of the oscillator-clock hypothesis. *Journal of Insect Physiology, 47*, 1097-1109.

Veerman, A. & Vaz Nunes, M. (1987). Analysis of the operation of the photoperiodic counter provides evidence for hourglass time measuremen in the spider mite *Tetranychus urticae*. *Journal of Comparative Physiology A, 160*, 421-430.

Veerman, A., Beekman; M. & Veenendaal, RL. (1988). Photoperiodic induction of diapause in the large white butterfly, *Pieris brassicae* : evidence for hourglass time measurement. *Journal of Insect Physiology, 34*, 1063-1069.

Volkovich, T. A. (1987). Role of light and temperature in controlling active development and diapause in *Chrysopa carnea* STEPH. (Neuroptera, Chrysopidae). 1. Photoperiodic reaction at constant and gradually changing day length. *Entomological Review, 66*, 1-10.

Volkovich, T. A. & Goryshin, N. I. (1982). Assessment of day-length in the threshold zone and the influence of the graduality factor on the photoperiodic reaction of *Pieris brassicae* L. (Lepidoptera, Pieridae). *Entomological Review, 61*, 15-24.

Wegis, M. C., Bradshaw, W. E., Davison, T. E. & Holzapfel, C. M. (1997). Rhythmic components of photoperiodic time measurement in the pitcher-plant mosquito, *Wyeomyia smithii*. *Oecologia, 110,* 32-39.

Wu, S. H., Yang, D., Lai, X. T. & Xue, F. S. (2006). Induction and termination of prepupal summer diapause in *Pseudopidorus fasciata* (Lepidoptera: Zygaenidae). *Journal of Insect Physiology, 52*, 1095-1104.

Xiao, H. J., Li, F., Wei, X. T. & Xue, F. S. (2008). A comparison of photoperiodic control of diapause between aestivation and hibernation in the cabbage butterfly *Pieris melete*. *Journal of Insect Physiology, 54*, 755-764.

Xue, F., Kallenborn, H. G. & Wei, H. Y. (1997). Summer and winter diapause in pupae of the cabbage butterfly, *Pieris melete* Ménétriés. *Journal of Insect Physiology, 43*, 701-707.

Zaslavski, V. A. (1972). Two-stage photoperiodic reaction as a starting point for the production of a model of the photoperiodic control of arthropod development. *Entomological Review, 51*, 133-145.

Zaslavski, V. A. (1988). *Insect Development: Photoperiodic and temperature control*. Springer, Berlin.

Zaslavski, V. A. & Fomenko, R. B. (1980). New data on the photoperiodic reaction of the knotgrass moth *Acronycta rumicis* L. (Lepidoptera, Noctuidae). *Entomological Review, 59*, 9-16.

Zaslavski, V. A. & Fomenko, R. B. (1983). Quantitative photoperiod perception in the aphid *Megoura viciae* BUCKZ. (Homoptera, Aphidae). *Entomological Review, 62*, 1-10.

Zaslavski, V. A. & Kats, T. S. (1988). Photoperiodic determination of sex in *Megoura viciae* (Homoptera, Aphididae). *Entomological Review, 67*, 32-40.

Zaslavski, V. A. (1995). Induction and deinduction in insect photoperiodism: experimental evidence and a model. *European Journal of Entomology, 92*, 401-413.

Zaslavski, V. A. (1996). Essentials of the environmental control of insect seasonality as reference point for comparative studies in other invertebrates. *Hybrobiologia, 320*, 123-130.

In: Biological Clocks: Effects on Behavior... ISBN: 978-1-60741-251-9
Editors: O. Salvenmoser et al. pp. 73-100

Chapter 3

RESPIRATION RHYTHMIC AND QUALITY OF SLEEP TO THE TOTAL WELLNESS AND DEVELOPMENT OF A CHILD

***Jong Yong Abdiel Foo*[1,2*] *and Stephen James Wilson*[3]**

[1]Division of Research, Singapore General Hospital, 31 Third Hospital Avenue, Bowyer Block A Level 3, Outram Road, Singapore 169608
[2]Bioprocessing Technology Institute, Agency for Science Technology and Research (A*STAR), 20 Biopolis Way, #06-01 Centros, Singapore 138668
[3]School of Information Technology and Electrical Engineering, University of Queensland, St Lucia Campus, Brisbane, Australia 4072

ABSTRACT

There is an increasing interest to improve the wellbeing of life especially amongst the middle to upper social class in the well-developed nations. Particularly, the study on quality sleep each night has refueled much attention to the overall wellness of these individuals. Studies have shown that disturbances during sleep not only can lead to mild irritations but also adverse events and life satisfaction as a whole. At the heart of the issue, the rhythmical functions of the body can be distorted when the occurrences of sleep disturbances become frequent. Sleep is now no longer viewed simply as a state of rest but it has specific and affirmative effects. Essentially, the brain is the

* Correspondence Author: Jong Y.A. Foo._Bioprocessing Technology Institute, Agency for Science Technology and Research (A*STAR), 20 Biopolis Way, #06-01 Centros, Singapore 138668. Email: abdielfoo@hotmail.com

prime beneficiary of sleep in order to maintain its cerebral capacities. Hormonal changes under the control of the sympathetic nervous system can increase blood flow to the muscles. This promotes the breakdown of stored nutrients into glucose in muscle tissue, thus providing energy available for the body to use. Monitoring the quality of sleep has stretched beyond just for individuals with susceptible sleep disorders but for many who want to maximize their daily lives. Compound with the fact that in many well-developed countries, birth rate has been low; caregivers are becoming more concerned in bestowing prime care for the long-term development and wellness of their children. However, limited research has been initiated to assess the quality of sleep in children until the recent decades. This can be due to the misconception or poor understanding of caregivers in pediatric sleep disorders. With much recent public knowledge in this domain, there is a paradigm shift in the mindset of caregivers. Thus, the once much-neglected yet specialized field of study has evolved since, from complex studies only conducted in fully equipped sleep laboratories to possibly simplified, miniature and portable ambulatory based studies. In this chapter, the comprehensive coverage of this topic will be presented together with a general perspective from a child, caregiver and healthcare provider.

INTRODUCTION

In all living organisms, there are variations in timing and duration of periodic or cyclic phenomena that occur throughout the life of the organism. Of which, one such phenomenon is related to sleep. Despite the fact that sleep is a known phenomenon since the beginning of time, most scientific research into this topic has only been conducted in the last century [1, 2]. Other than the experience of being "out-of-touch" with the environment, this relative absence of mental activities during sleep has led to the common belief that the brain is being "turned off" [1, 3]. On the contrary, recent studies have shown that sleep is actually a special activity of the brain that is controlled by elaborate and precise mechanisms. Sleep is not simply a state of rest, but it has very specific and affirmative functions. In particular, the brain is the prime beneficiary of sleep in order to maintain its cerebral capacities [2, 4-5]. Furthermore, hormonal changes under the control of the sympathetic nervous system will increase blood flow to the muscles. This promotes the breakdown of stored nutrients into glucose in muscle tissue, making energy available for the body to use [6]. More specifically, sleep not only sub-serves complex inter-related physiological processes but also is conspicuously critical to a developing child. Studies have shown the importance of sleep to endocrine and respiratory physiology of a child [5, 7]. For example, the association of growth hormone secretion with sleep in pre-pubertal children and the impact of reduced nocturnal respiratory function in chronic respiratory and neuromuscular diseases.

The degree to which the human body can function effectively will largely depend on the duration and quality of sleep. Ideally, there is a biological rhythm of routine cycles in wakefulness and sleep [3, 6]. However, it is found that there are disturbances to the sleep cycle. These disturbances not only can cause daytime sleepiness and affect alertness, but also may induce changes in the sleep patterns [1-2, 6]. Moreover, studies have shown that the body is more prone to illnesses with unfulfilling sleep. Thus, the consequences of disturbed sleep can range from annoying to life threatening episodes that can include depression, learning difficulties and falling asleep while driving. Compared to the adults, quality sleep is even more prevailing to children and infants. The accumulative detrimental effects of disturbed sleep have shown to affect the neurological, physiological and sociological development of these young ones [5, 7-9].

There have been notable interests in the middle to upper social class of many well-developed nations to maintain the wellbeing of life. The study on quality sleep each night that has been dominant for a long time was due to the much-neglected public knowledge in this domain. With the recent paradigm shift in the mindset, this has refueled much attention to the contributions of quality of sleep to the overall wellness of life. To many, sleep is now no longer viewed simply as a part of the daily routine. Monitoring the quality of sleep has stretched beyond just for individuals with susceptible sleep disorders but for those who want to maximize their daily lives. The fact that total fertility rate (or the average number of children that would be born to a woman over her lifetime) has been low in industrialized countries like Italy and Japan as well as in developed countries like Hong Kong and Singapore [10] as shown in Figure 1. Particularly, the value of replacement-level fertility is generally assumed to be around an average of 2.1 lifetime births per woman. Failure to maintain this level can result infertility levels that are below replacement and lead to long-run population decline [11]. Therefore, the sustainability of the retirement income, health and aged care systems with the given increase in dependency ratios has been becoming increasingly a concern to the governments. Besides the possible social impact and/or imbalance, caregivers in these countries are also becoming more concerned in bestowing prime care for the long-term development and wellness of their children. In view of this, the wellness of a child then becomes critically important to both the caregivers and to be considered as a national social issue.

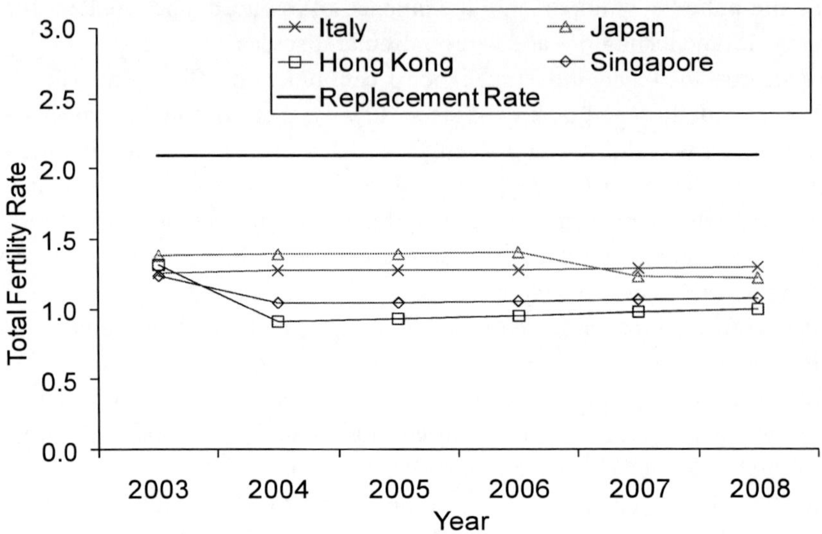

Figure 1. The total fertility rate in many well-developed countries is below the recommended replacement rate. This has complex impact on the social dependency ratio and perception on the total wellness of each child. Source: CIA World Factbook, Washington, DC [10].

SLEEP AND BREATHING

Breathing is an essential physiological process where it involves gases exchange. Particularly, the body acquires oxygen from the atmosphere while it releases carbon dioxide, the major end product of oxidative metabolism. Inspiration and expiration during normal breathing cycles can achieve these biological requirements of the body. Anatomically, the whole breathing process involves the passage from the nares at the nose to the alveolus in the lungs or commonly known as the airways [12-14]. Therefore, it is important that this pathway is unobstructed at all times for this process to occur. However, there are incidences when obstructions or abnormalities do occur. It is further easier for an adult to voice out such occurrences. However, for younger children and infants, they may not know how to explain it or draw the appropriate attention. This is often misunderstood by caregivers in that the child is throwing a tantrum and refusing to go to bed [7-9, 15].

Daytime occurrence of such abnormalities can be easily noticed for all age groups as this often leads to coughing in their attempts to overcome the obstruction. However, there may not be any prominent evidences of such events during sleep unless they are closely monitored. If it were to be left unnoticed for long period of time, there can be adverse and permanent effects on the child [7-9]. Moreover, death can even occur in infants in severe cases and is generally termed as sudden infant death syndrome or SIDS in short [16]. For these reasons, there is a need for continual observation in a control environment. Sleep studies are established where the nominal breathing patterns of individuals are monitored throughout their sleep, including children and infants. The recordings from these overnight sleep studies can be useful to help identify any breathing abnormality in these young ones. Appropriate advice and/or treatment can then be administrated to improve their situation and thereby wellness.

Sleep Disturbances

It is found that there are a number of reasons that can cause disturbances in sleep. Some of them are factors extrinsic to the sleep process like sleep postures, emotional issues and the surrounding environment [1-2, 6]. These can occur randomly and would not exist for an extended period of time and can generally be minimized with suitable guidance or preventive measure. However, there are other causes that warrant concerns for appropriate investigation. These can include a broad range of factors like dysfunctional sleep mechanisms, abnormalities in physiological functions during sleep and the malfunction of the biological clock. These intrinsic causes are more critical and may require professional medical assistance. Of great interest to sleep scientists globally is sleep apnea; this is one of the most common sleep-related breathing disorders in the general population that can induce disturbed sleep. The occurrence of sleep apnea cut across all age, gender, culture, geographical location and ethnicity [4, 7-8].

The Human Upper Airway

Understanding the physiology and anatomy of the human upper airway provides better insights to the occurrences and issues relating to sleep-disordered breathing. The upper respiratory tract in the human body performs a wide range of respiratory and non-respiratory functions [17-18]. The primary focus in this context is geared towards the respiratory function of the upper airway motor elements that influence breathing. There are growing interests on the role of the upper airway especially during sleep. In a large part, this has come about with the increased recognition of the entity of obstructive sleep-disordered breathing. Patency of the upper airway depends on transmural pressure that is the difference between the negative intra-thoracic pressure caused by inspiratory efforts and the positive dilating pressure from the upper airway musculature [19].

Structure of the Upper Airway

The respiratory role of the upper airway musculature in humans is the maintenance rather than production of airflow. Upper airway resistance is highly variable even during quiet and normal breathing. The chief sites of variable resistance are the nose, mouth, pharynx and larynx; the anatomical location of these sites is as shown in Figure 2 [13-14, 19]. These sites may be viewed anatomically different but it is important to note that they work as a unit in many circumstances and are highly interdependent. Particularly, the total structure forms a passage for air movement flowing from the nose to the lungs and also participates in other physiological functions. The various properties of the upper airway are a compromise between different physiological functions that maintain its patency during breathing and its closure during swallowing.

The nose or nasal airway accounts for about half of the total respiratory resistance in normal humans at rest. Resistance to airflow is highly variable both within and among subjects. During rapid respiratory efforts such as sniffing, the external nares supported only by the soft tissues of the nose are subjected to dynamic compression. However, the resistance to airflow does not increase beyond that experienced at a negative intrathoracic pressure of approximately 8cmH_2O. An important feature of the nasal airway is that it can be partially bypassed by mouth breathing. The conversion between nasal and oral breathing during events of increased inspiratory efforts is unclear; whether it is due to reflex origin or learned behavior. Interestingly, total nasal resistance does not change on transition from wakefulness to sleep [18].

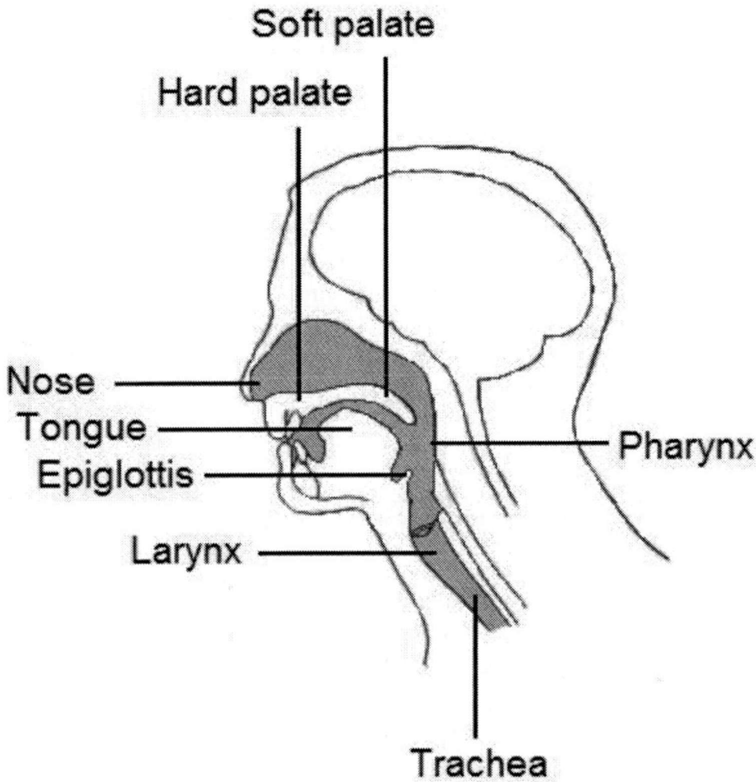

Figure 2. Sagittal view of the human upper airway where all the stated anatomical sites have contributing roles in the patency of the airway.

The respiratory behavior of the pharynx and the muscles that influence it can affect the airflow. The soft tissues of the pharyngeal walls are quite compliant and these tissues provide this portion of the airway with rather fluid anatomical relations. Airflow resistance can change rapidly, both in response to glossal and pharyngeal muscle activities. Moreover, this can change passively as a consequence of transmural pressure changes caused by breathing movements of the diaphragm and other muscles of the ventilatory pump. Similar to the case with the external nares, the pharyngeal airway is subjected to inspiratory narrowing or occlusion by dynamic compression. In relation to intrathoracic airways, an increased nasal resistance can aggravate this problem. During the onset of an obstructive sleep apnea, the tongue or other surrounding muscles may relax against the pharyngeal walls limiting the airflow [2, 20].

Although the larynx is capable of varied and complex movements in activities such as swallowing, coughing and speaking, the variations in laryngeal resistance during normal breathing take place largely at the vocal cords. To understand the laryngeal respiratory movements involves the consideration of the functional anatomy of the larynx and its intrinsic as well as extrinsic muscles. Respiratory movements of the vocal cords are accomplished by the rotation of the arytenoids about their ligamentous attachments to the cricoid cartilage. However, these respiratory movements are not determined exclusively by activities of the intrinsic laryngeal muscles. Extrinsic muscles extending from the cricoid and thyroid cartilages to other structures form a suspensory sling for the larynx. They exhibit some phasic activity with the breathing cycle, particularly in the hyperpneic states [17, 19].

Factors that Affect the Upper Airway

The physiology of breathing in the human body is a complex process. In order to comprehend this process, static and dynamic factors that contribute to the net patency of the upper airway need to be considered. The former factors include the effect of the gravitational forces on the compliant structural support, shape and size of the airway. Thereby, the airway is then dependent on the position of soft tissue structures like soft palate, tongue and walls of the pharynx. In supine posture, the tongue and soft palate have shown to move posteriorly thus reducing the pharyngeal area and increasing collapsibility [12, 17]. Clinical observations have demonstrated that snoring is more prominent in this posture [16, 18]. The following factors also comprise the anatomical structure of the upper airway like fats and the muscles surrounding it. Even though the airway size stays fairly constant during inspiration, resistance from these fixed structures can limit the airflow. Dynamic factors consist of the respiratory driving pressure across the region susceptible to collapse that is determined by the amount of negative intrathoracic inspiratory pressure. In particular, the intrinsic properties of the airway walls have an important role in its patency. Furthermore, the neural inputs to the upper airway like chemoreceptors and mechanoreceptors can also dictate the behavior of the dilating musculature [21].

Sleep Apnea and Hypopnea

Sleep apnea is a disorder that consists of brief periods throughout the night in which breathing can stop. Generally, people with sleep apnea do not get enough oxygen in their sleep. There are two possible sleep-related processes that can contribute to the occurrence of sleep apnea. Firstly, the level of brain activation declines at sleep onset, particularly in the brainstem reticular formation that contains the neural oscillator that in turn controls breathing. The other process contributing to sleep apnea is the decline in the muscle tone throughout the body during sleep. This includes the muscle tone of the tongue, back of the mouth, throat and pharynx [8-9, 16]. In order to understand the pathophysiology of these two processes in sleep apnea, they can be broadly categorized into three different types namely obstructive, central and mixed. Besides apneas, periods when breathing efforts are reduced by approximately 50% or more while displaying similar characteristics as apnea are called hypopnea [7-9]. Likewise, they can be generally categorized as obstructive, central or mixed in nature.

Central Respiratory Events

Central respiratory events are considered here as the occurrences of either central apneas or hypopneas. These respiratory events are generally classified as a neurological condition. Compared to its counterpart, the obstructive respiratory events, apneic terminating arousal is less intense and cardiac arrhythmia is less frequent in the central oriented ones [22]. However, central respiratory events still cause the cessation of all respiratory efforts during sleep, usually with a decrease in arterial blood oxygen saturation or in short, SaO_2. These respiratory events arise as the central drive to respiratory muscles and airflow cease even when the summations of respiratory stimuli fall below a threshold level that is required to generate ventilation as given in Figure 3 [23]. This may be caused by the insensitivity of the chemoreceptors to the onset of such events [24] and there is generally no apparent physiological response even though hypoxemia can be observed. The onset of obstructive respiratory events may end with arousals to initiate of airflow; however arousals are usually absent at the termination of the central oriented ones [22- 23].

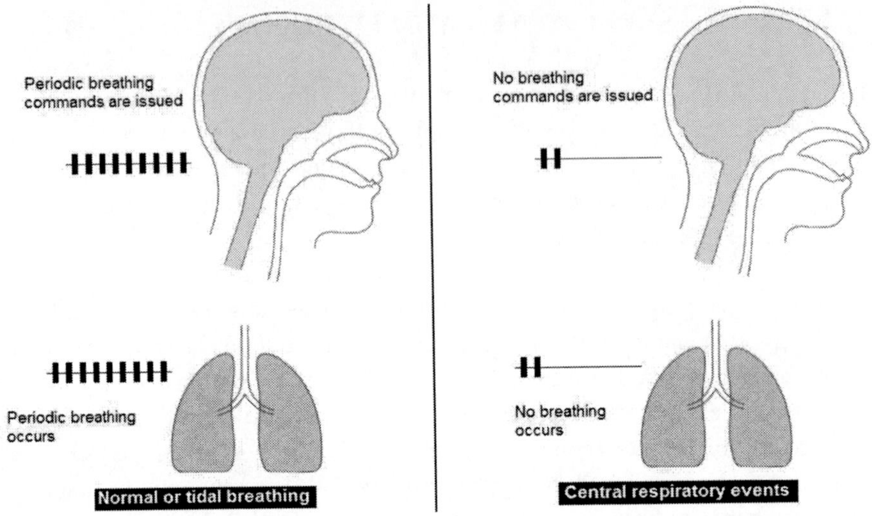

Figure 3. As the respiratory center of the brainstem is a part of the reticular formation, commands to breathe are frequent and regular during waking when reticular neuronal activity is high (left-side). During sleep, the automatic brain commands to breathe become less frequent and may stop altogether (right-side). This process is called central sleep apnea and may become pathological if it becomes prolonged and frequent. When this occurs, it reduces oxygen concentration in the blood and can cause numerous awakenings.

Previous findings have suggested that a number of mechanisms have been identified by which central respiratory drive can be transiently inhibited [7, 24-25]. These mechanisms include defects in the respiratory control system or muscles, transient instabilities in respiratory drive and reflex inhibition of the central respiratory drive. During sleep, breathing is mainly regulated by the metabolic respiratory control system that receives input from the peripheral and central chemoreceptors as well as mechanical ventilatory events in the airways and lungs. Typically, respiratory system with this disorder may be compensated and unknown to the individual when awake. However, it becomes fully manifested during sleep. Secondly, if the central nervous system fluctuates between the "awake" and "asleep" state, there is a difference in the stimuli threshold required to generate the respiratory rhythm. During these transient periods, fluctuations in the central respiratory drive can occur. Lastly, there are reflexes that are capable of transient abolition of the central respiratory drive. These include the pulmonary inflation reflex and inhibitory reflexes arising in the upper airway like the muscles surrounding it [22-25].

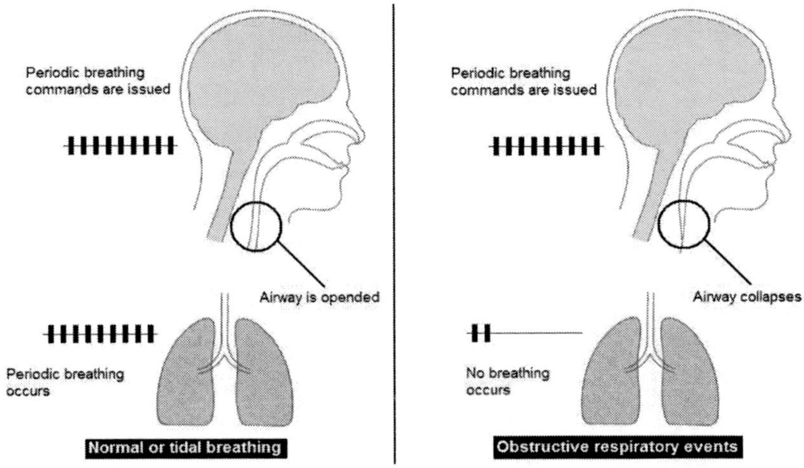

Figure 4. The reticular neurons control the muscles that hold the airway open in awaked individuals (left-side). However, these neurons diminish their output during sleep (right-side). Due to this peripheral mechanism, the airway may completely collapse and this is termed as obstructive sleep apnea.

Obstructive Respiratory Events

Similar to the central respiratory events, obstructive respiratory events are defined as episodes of obstructive apneas and hypopneas. These obstructive respiratory events are generally considered to be anatomical or functional conditions [9, 26]. Since the main difference between apneas and hypopneas is their relative degree of ventilatory, only the pathophysiology of obstructive apnea is examined in this context. Obstructive sleep apnea is characterized by repetitive pauses in breathing during sleep due to the obstruction and/or collapse of the upper airway. Figure 4 shows that diaphragmatic respiratory effort continues during the episodes of obstructive sleep apnea and is generally followed by an awakening to breathe [1-2, 22]. Interestingly, obstructive sleep apnea is usually regarded to be clinically relevant if this happens more than five times per sleeping hour. This can be due to the fact that occurrences of obstructive sleep apnea in any healthy individual are also common [1-2, 6]. During each obstruction, it can cause a fall in the blood oxygen level, which is followed by an increase in the sympathetic nervous activity. This can then lead to an increase in heart rate and blood pressure, thereby causing possible arousals [27-28]. Generally, individuals with obstructive sleep apnea can experience snoring, laborious breathing or profuse sweating during sleep. In short, the flow of air is restricted and cannot be ventilated through the lung tissues even though normal breathing efforts continue.

Besides the effects experienced by adults as detailed in the previous paragraph, younger children and infants with obstructive sleep apnea may also undergo breath-holding spells when they are awake. They may experience difficulty in breathing and have poor swallowing coordination during meal times. In severe cases, visible effects can be seen in younger children and infants like delayed growth, abnormal weight and recurring ear infection [16, 29]. However, the degree of these effects greatly depends on the child. The critical thing is then to detect these abnormalities early so that proper treatments can be advised and administrated. During an obstructive sleep apnea episode, cessation to breathe continues until that individual wakes up from it. The arousal instantly increases the activities of the muscle in the tongue and the throat, when this occurs the airway will be enlarged relatively. The individual will then be able to breathe and fill the lungs with the required oxygen once again. This vicious cycle may be repeated numerous times in a night while the individual is unaware that it is happening throughout the sleep.

Anatomically, obstruction of the upper airway can occur in people who either have a narrow airway or those whose muscles around the throat behave abnormally. In the latter case, the tongue, the pharynx and other surrounding soft tissues relax against the throat during sleep [2, 20]. This leads to a partially blocked upper airway and thereby restricting the flow of air in the body. Consequently, the body is unable to inhale sufficient oxygen and exhale excessive carbon dioxide. In particular, this causes a low level of oxygen or hypoxemia and high level of carbon dioxide or hypercarbia in the bloodstream [16]. This alerts the brain to coordinate remedial action to overcome the upper airway obstruction. The increase in blood pressure or heart rate is the remedial response from the brain and this may lead to snoring or laborious breathing. Unknown to the person, the frequency and duration of these breathing pauses prevent the body from getting a restorative and normal sleep.

There can be four major predisposing factors for the upper airway obstruction. Firstly, anatomical narrowing can occur when a smaller cross-sectional area of the upper airway has a decreased ability to maintain upper airway patency. Secondly, an abnormal mechanical linkage between airway dilating muscles and airway walls can diminish the ability of pharyngeal wall to stiffen the airway, thus increasing collapsibility. Thirdly, muscle weakness can contribute to an imbalance of pressures faced by the upper airway during inspiration and expiration. Lastly, abnormal neural regulation like a degree of blunting in central chemosensitivity can affect the patency of the upper airway [8, 30].

Table 1. Some of the common detrimental effects children with sleep-disordered breathing can experience

PHYSIOLOGICAL EFFECTS
Hypoxemia, hypercapnia, profuse sweating, premature ventricular or atrial contractions, delayed growth, easy prey to disease, right ventricular dysfunction, abnormal weight, recurring ear infection
NEUROLOGICAL EFFECTS
Seizure disorder, developmental delay, learning problems, poor academic performance
SOCIAL EFFECTS
Aggressiveness, hyperactivity, pathological withdrawal, daytime sleepiness

Mixed Respiratory Events

As its name suggests, mixed respiratory event is a combination of the central and obstructive respiratory event. An episode of mixed respiratory event usually starts with a central component and then eventually becomes obstructive in nature [1, 22]. Generally, it is believed that the central component of the respiratory events becomes less prominent once the obstructive counterpart is treated. The separation of respiratory events into central and obstructive, although useful in conceptualizing the pathophysiology can be somewhat misleading. All respiratory events at least in part, are central in origin. Neural control of breathing during sleep not only initiates diaphragmatic activity, but also is also responsible for the upper airway muscle activity integral to the maintenance of its patency [16]. The physiological basis for the definition of mixed respiratory events at this stage is perhaps the most questionable. Knowledge of whether the upper airway is closed or open during such events is important if its definition is based on breathing effort or physiological effect [31]. The specific definition of these events would require a comprehensive understanding of the key characteristics of the various respiratory events but this would not cover within the present context.

Obstructed Upper Airway in Children

The detrimental effects as described in Table 1 can be considered to be clinical complications of obstructive sleep apnea in children. Some of these are specific as manifestation of obstructive sleep apnea whereas others may be seen in a variety of conditions that can be contributed by obstructive sleep apnea [16]. These findings can be categorized as effects of obstructed breathing for both asleep and awake state. Although the frequency of these occurring has not been studied in a controlled fashion, it is useful to understand the possible detrimental effects observed in children with obstructive sleep apnea syndrome. Table 1 gives some of these effects which are tabulated as physiological, neurological and social accordingly.

INSTRUMENTATIONS OF SLEEP STUDIES

Accurate measurement of respiratory efforts and identifying the occurrence of respiratory events are the key determinants in respiratory sleep studies [38]. The measurement of these physiological events occurring in the body during sleep has not been possible until the discovery of electroencephalography (also known as EEG) in 1875 and the discovery actually lays the foundation for the present sleep study [4, 6, 32-34]. However, there is no single or simple measurement available that is able to ascertain the breathing patterns of individuals during sleep. Specifically, more objective information is necessary to assess the quality of sleep. Current diagnostic sleep studies usually have the following electrophysiological measures as the minimum requirement; monitoring brain waves, eye movements and muscle activities [1]. The current best determiner is to conduct a nocturnal diagnostic sleep study and the widely used technique in a typical pediatric sleep laboratory is the polysomnography (or PSG in short) [7-9]. With polysomnography, the presence, extent and depth of the sleep-disordered breathing can therefore be monitored simultaneously. Thereafter, the findings from such comprehensive monitoring can lead to appropriate medical advices and proper treatments to be administered.

Diagnostic Polysomnography

Diagnostic polysomnography is a collective process of monitoring and recording the vital physiological signals of an individual during sleep. With these data, identifying and analyzing abnormal respiratory events throughout the sleep process are then possible. In particular, polysomnography provides an objective measure of the number of apneas and hypopneas that occur during sleep. It also gives a measure of the consequences of breathing disturbance like the degree and duration of hypoventilation, oxygen desaturation, cardiac arrhythmias and sleep disruption [9, 16]. It is generally recommended that studies be performed overnight but daytime monitoring is not uncommon like sleep latency tests [34]. Despite its wide use, the polysomnography setup can vary in techniques from one sleep laboratory to the other including some of the measured parameters [39]. However, understanding the explicit differences in these setups is not of grave importance. In Figure 5, it illustrates the more commonly used electrophysiological measurements in most pediatric polysomnography study while in Table 2, it details the main parameters these measurements monitor during respiratory sleep study.

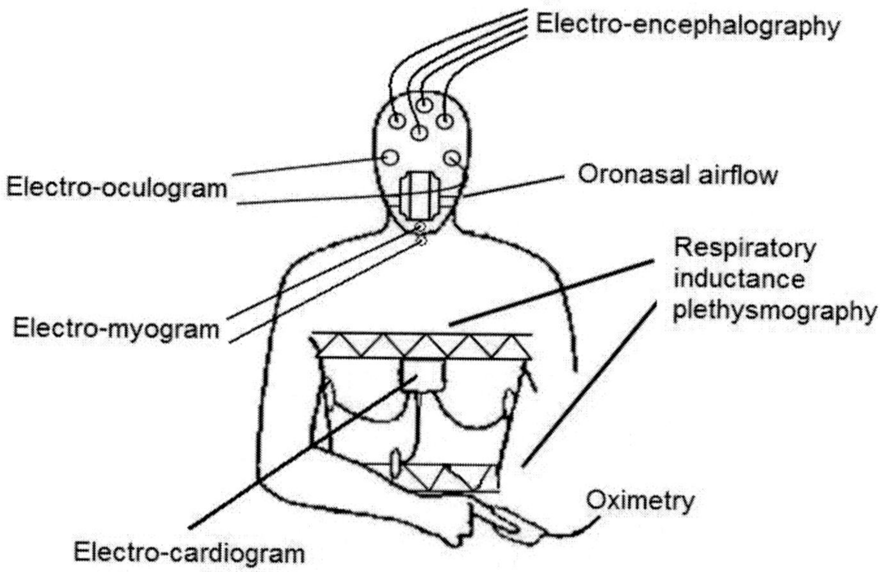

Figure 5. General electrophysiological measures used in overnight polysomnography studies in the investigation of sleep related breathing disorders

Based on the standard polysomnography scorings, an obstructive sleep apnea episode is defined as a complete cessation of airflow with the presence of chest and/or abdominal wall movement while an obstructive hypopnea has similar characteristics except that airflow is reduced by at least 50% from its baseline [42-43]. From the same references, central hypopnea events are defined as a reduction of 50% or more in airflow proportional to the decrease in chest wall movement while central sleep apnea is the complete cessation in both airflow and respiratory drive. The two commonly used techniques in polysomnography to assess the occurrences of respiratory events are the respiratory inductance plethysmography and oronasal airflow measurement. The former technique is usually used to monitor the respiratory efforts through the synchronization of chest and abdominal movements while the latter can be determined by either changes in temperature or pressure during inspiration and expiration. Figure 6 and 7 shows the typical polysomnographic recordings during events of central respiratory events and obstructive respiratory events in a sleeping child respectively.

Table 2. The general measures and uses of electrophysiological measures in the polysomnography protocol to investigate sleep-disordered breathing

Parameters	What it generally measures	Use in sleep studies
Electro-encephalography (EEG)	Recordings of electrical activity generated by the brain. Its characteristic frequency and amplitude patterns are generally analyzed.	Monitor the transitions from wakefulness to sleep. Also define the sleep stages.
Electro-oculogram (EOG)	Recordings of movements of eyes	Differentiate rapid eye movement (or known as REM) from non rapid eye movement (or NREM)
Electro-myogram (EMG)	Recordings of electrical activity of the muscles	Identify arousals from wakeful state
Electro-cardiogram (ECG)	Measure of electrical activity of heart	Detect apneic events and oxygen desaturation that can cause dramatic arrhythmias
Respiratory inductance plethysmography (RIP)	Determine chest wall and abdominal movements during breathing	Identify presence and extent of respiratory events in conjunction of airflow
Airflow	Monitor amount of air in and out of the airways	Determine presence and extent of respiratory events in conjunction of respiratory inductance plethysmography
Arterial blood oxygen saturation (SaO_2)	Measure arterial oxygen saturation	Assess the amount of arterial oxyhemoglobin de-saturation during respiratory events. This measure is usually derived from an oximetry.

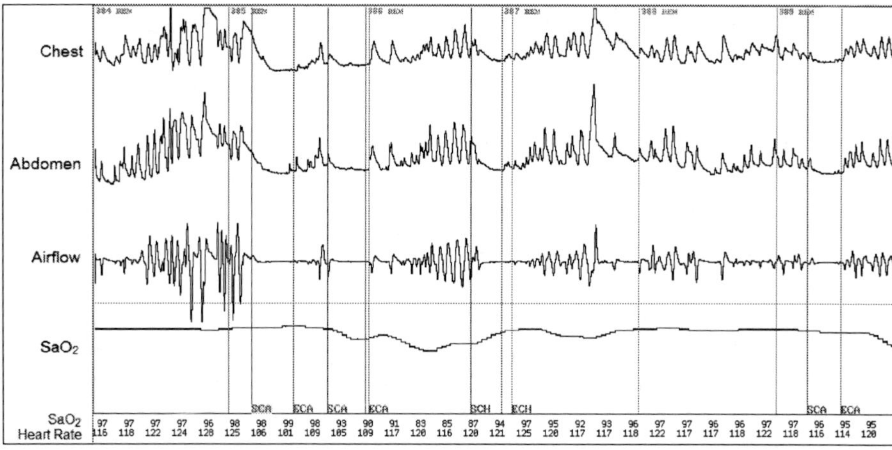

Figure 6. Mixture of central apneas and hypopneas in a standard nocturnal polysomnography study where SCA and ECA denote the start and end of central apnea. During the central mediated events, it can also be seen that the arterial blood oxygen saturation (or SaO_2) value of the child deteriorates with the progression of the prolonged central events.

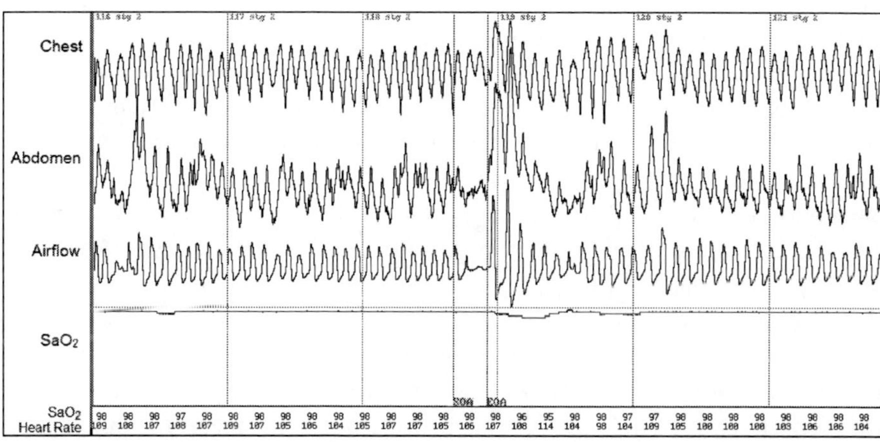

Figure 7. An obstructive event observed in this polysomnography study there is a transient and gradual decrease in SaO_2 (or arterial blood oxygen saturation). Legend: SOA and EOA denote the start and end of obstructive apnea respectively.

Epidemiology and Natural History of Diseases

There are evidences that suggest sleep disturbance can be common in the general pediatric population [7-9, 33-34]. Findings from both clinical observations and experimental investigations have suggested that sleep disturbances often can have compelling effects. Due to the limited recognition and referral cases for medical attention, the exact accuracy and intensity of sleep disorder across all the age groups is unclear. Much greater awareness and diagnostic availability to the general population are needed before reliable numbers can be properly tracked [2, 4, 16]. At present, it can be reasonable to assume that the quoted figures for incidences of sleep disorders in children are generally underestimated [30, 34-35]. Specifically, there are not many extensive epidemiological studies of obstructive sleep apnea syndrome conducted on children. However, the incidence rate of children in the very limited studies done in the few countries would be used as a point of discussion within this context.

In a study by Blunden *et al* [35], they have used snoring as a marker for sleep-disordered breathing in children. This approach is consistent with other likewise studies to determine the incidence of sleep-disordered breathing in children [7-8]. According to a study conducted in the United Kingdom, it suggests a prevalence of obstructive sleep apnea syndrome in 4 to 5 years old to be between an incidence rate of 1.75% and 2.25%. Conversely, the prevalence of snoring in the general unselected population has been estimated to be about 20% [16]. From these percent figures, it can be deduced that the rate of snoring increases dramatically in the general population with age. Regular snoring is a universal finding in children with symptomatic obstructive sleep apnea syndrome [35]. Other studies have also suggested that snoring in early life may contribute to the eventual development of the same symptoms later in life [9].

In some studies conducted in the United States, obstructive sleep apnea syndrome occurs in at least 8% of school age children [30, 34-35] and the corresponding incidence in childhood is also estimated at 1 to 3% [32]. Likewise, snoring is used as the symptom to classify individuals with obstructive sleep apnea syndrome in these studies. Based on the 1988 census numbers, there were approximately 3.6 million children who were 4 years of age in the United States. If the conservative approach of 1% obstructive sleep apnea syndrome prevalence estimates were to be adopted, there would have been approximately 36,000 children with obstructive sleep apnea syndrome per year for just one age group [16]. This is probably just the tip of the iceberg. There would possibly be about half a million of children with obstructive sleep apnea syndrome if all other age groups of children were taken into considerations [35]. It is worth noting that these projections are not mere wild speculations. Clinical findings have shown that young individuals, who have been snoring since childhood, account for approximately 70% of adult patients with symptomatic obstructive sleep apnea syndrome [16]. Therefore, it is vital that professional medical care is provided to these children to improve their long-term wellness and developments [42-43].

The diagnostic polysomnography study is not the viable solution to identify sleep disturbances in the general population. Some key factors to consider are the operational costs and the complexity of the essential measurements to be managed [44-45]. It is then only realistic that the polysomnographic studies are conducted in well-equipped sleep laboratory. However, this approach has inevitably created the presence of a long list of patients waiting for their sleep studies to be administrated [45-46]. Thus, the low availability of the sleep laboratory relative to the number of patients and the burden of its associated costs has made polysomnography unreadily accessible to the general public [44, 47]. A study conducted in Paris has showed that the diagnosis of obstructive sleep apnea syndrome in 25% of their patients can be delayed up to 10 years [44]. In short, polysomnography by itself may be ineffective to handle the pediatric referrals for sleep-disordered breathing investigation. Based on the current circumstances, there is a clear need for simpler techniques to screen children who are susceptible to obstructive sleep apnea syndrome before a full diagnostic polysomnography study is conducted on them [48]. Over the recent years, substantial interest has been focused on portable devices that have the ability to record a combination of cardio-respiratory parameters for ambulatory or home use [47, 49]. This should not only accelerate the process of sleep disturbance diagnosis, but also help reduce the overall associated costs [44]. Furthermore, an ambulatory setup can probably provide a more realistic appraisal of sleep pathology than in the laboratory settings [50-51]. However, these recordings need to be reasonably useful for clinical assessments of respiratory events in children. Few techniques have proven their capability to assess the respiratory variables for remote or home monitoring with great accuracy [52-53].

CURRENT DILEMMAS IN RESPIRATORY SLEEP STUDIES

Besides the cost and availability of polysomnography to the general public, issues related to respiratory sleep study needs to be viewed from a multi-faceted angle [34, 46-47]. As a point of discussion, the referral after verifications from their general practitioner for children with suspected obstructive sleep apnea syndrome to better-equipped sleep laboratory in a nearby major city is used.

Particularly, the referrals from the outback of the Queensland State to the Mater's Misericordiae Children Hospital located in Brisbane of Australia are used as an example. On the surface, this arrangement seems to address the need to diagnose the pediatric population of this geographical region adequately. However, there are gaps to be plugged before this seemingly ideal approach becomes a reality.

A Healthcare Provider's Perspective

The quality and effectiveness of healthcare providers largely depend on their ratio to the general population. A high ratio is always desirable but this will inevitably incur higher associated costs and thereby reduce the corresponding efficiency. With the same token, the number of medical equipment and required technical expertise are limited in the same manner. Thus, it can become a socioeconomic issue where healthcare providers are expected to fulfill their responsibilities promptly and yet stay cost effective. Healthcare providers are then expected to learn to identify the crucial cases and prioritize them accordingly. In the investigation of sleep-disordered breathing, pediatric respiratory staff is expected to adopt similar professionalism and mindset. However, there is a gap between the clinical needs and provision of services for sleep disorders medicine, especially in children [43-44]. Unfortunately, there are limited features on sleep disorders for the medical, nursing and allied health professionals during healthcare teachings or trainings [34]. It is observed that few medical professionals address the sleep problems faced by a child and may possibly also offer ineffective treatments that may aggravate their conditions. In addition, there are very limited established sleep laboratory providing appropriate healthcare services to children with sleep disorders [7-9, 40].

In the scenario where all the children with suspected obstructive sleep apnea syndrome in the Queensland State are referred to the Mater Misericordiae Children's Hospital in Brisbane, this can unknowingly increase the number of sleep studies conducted with a negative diagnosis and rendering a poor utilization of these limited sleep laboratory. Thus, it may not be worthwhile conducting nightlong sleep studies on all referral cases as this may slow down the process of attending to genuine cases [44, 54]. In simple words, the corresponding workloads are unnecessary higher due to the extra nocturnal studies that are performed. In order to conduct overnight sleep studies, specialized medical staffs are required. Not only does it require expertise skills, but the preparation and setting up for each study can be tedious [34, 55]. With high overheads to manage and lower financial returns (when compared to other medical specialties), insufficient efforts are put in for sleep laboratory across most healthcare institutions. In this regard, each existing sleep laboratory not only needs to provide services to children in their vicinity, but also possibly at a much larger geographic scale as with the Mater Misericordiae Children's Hospital. Frequently, proper logistic arrangement, prioritizing and scheduling of these patients then become a mounting task to ensure that every sleep laboratory is optimally utilized [44].

A Caregiver's Perspective

Due to its poor recognition, the general population including parents or caregivers may have a misconception of sleep disorders [7, 36]. There is a widespread dogma that everyone experiences difficulties with their sleep sometimes in their lifetime. The difficulty in falling asleep is usually seen as temporary and is often regarded more as a nuisance rather than anything serious to warrant special attentions [53]. In regards to a child's sleep problems, caregivers normally deem that as the child refusing to go to bed and staying awake for attention [9, 34-35]. In addition, some caregivers adopt the belief that a child's sleep problem is part of the mixed blessings of parenthood or as something to be endured until the matter improves naturally. Regardless of the period of the child's sleep difficulty, most caregivers do not see the need to seek medical advices. To many, the perception is that these sleep problems tend to decrease with age. The fact of the matter is many of these children do not outgrow their sleep problems as is commonly assumed [7, 34]. There are also some misconceptions that sleep problems do not merit a great deal of professional medical attentions when compared to other medical conditions. Furthermore, there are beliefs that sleep problems is a result of a child not having learnt the appropriate sleep habits or patterns. Simply, the opinion of caregivers on a child's sleep problems is very critical because that will determine the caregivers' approach towards a child's sleep behavior and whether appropriate helps will be sought for the child's sleeping condition [53].

A Child's Perspective

Unknown to caregivers, sleeping alone in a laboratory bed can make some children feel afraid, uncomfortable or anxious [7, 34]. Sadly, the overnight polysomnographic studies conducted in sleep laboratory can be an external stimulus that induces sleep disturbances to a child, especially on the first night of the study [56-57]. Firstly, the child is put in an environment or setting that is unfamiliar to the child. This can induce unnecessary stress to the child. Secondly, the feeling of being under continuous observation does not appeal to most children and this can indirectly affect the quality of sleep [56-57]. Particularly, prolonged sleep latency, shorter total sleep time, reduced sleep efficiency and more time awake during the night are some of the effects a child may experience in the laboratory study [58-59]. In short, these generalized consequences are commonly known as the first-night effect [56]. No matter how much effort is put into creating a friendly or relaxed environment in the sleep laboratory, it still cannot replace the familiar and conducive setting that the child is comfortable with just like in the bed at home [56-59]. Clinical findings have suggested that data obtained from the first night of the study should be discarded and to be excluded in the assessment of one's quality of sleep [56]. This would simply mean that an additional night in the sleep laboratory is required and this can cause inconvenience to some especially caregivers who are in the workforce. There are even perception from some children that sleep laboratory have frightening sights, sounds and smells [34].

There are recommendations that sleep laboratory should include facilities for caregivers to stay overnight with the child during their sleep study. The inclusion of such facilities can only help to a certain extent but may not be sufficient to eliminate the common negative effects of spending a night in the sleep laboratory [34, 59]. In general, a standard sleep study requires a number of measuring electrodes to be put on a child. This not only causes much discomfort to the child [57] but the electrodes can also be inadvertently entangled during the night. Figure 8 shows the some of the standard polysomnography electrodes that need to be placed on a child through the nocturnal study. There are numerous efforts put in to develop techniques to simplify the sleep study monitoring process [46-55]. However, in order for remote monitoring to be possible, simple yet robust measuring techniques to monitor sleep-disordered breathing in children needs to be developed first.

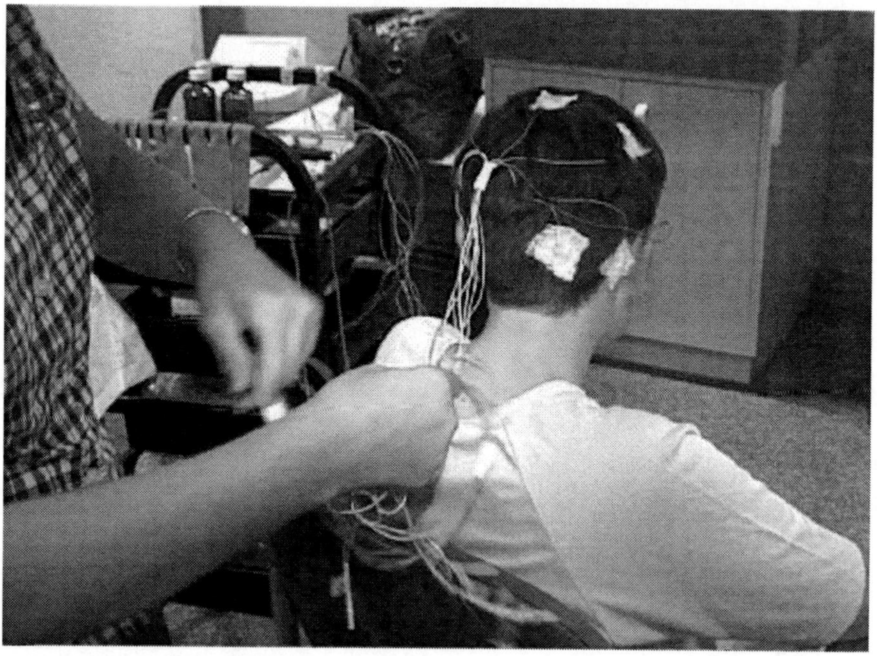

Figure 8. Standard electrodes are being applied on the child before commencement of the overnight polysomnography study

CONCLUSIONS

The interest to improve the wellness in life has fuelled much growth in the biomedical research arena. Particularly, the evaluation of one's quality of sleep each night has been attracting increasing attention Coupled with the fact that in many well-developed countries, birth rate has not been less than ideal. Caregivers are becoming more concerned in bestowing prime care for the lifelong development and wellness of their child. Respiratory sleep studies have been taking an evolutional turn in that more is now expected to meet the increasing need for better yet simple assessment techniques. Few techniques have reached the stage to be field tested for ambulatory or home monitoring. The promotion of these techniques is not an attempt to replace the standard polysomnographic studies but it is hoped that these techniques can provide an adequate or provisional indication about the degree of sleep disturbances in a child. It is clear that a simplified cardio-respiratory technique is preferred as caregivers and the child are more susceptible to such methodology of measurement. Such preference can be due to the fact that fewer apparatus are required to be placed on the child. Hence, it would be more manageable for the caregivers in making home monitoring a reality. Thereby, not only the waiting list for sleep disorders investigations can be shortened but also the overall wellness of the precious mini biological clocks will be known at their caregivers' fingertip.

ACKNOWLEDGMENT

The authors acknowledge the contributions and support towards the work described herein from the Department of Respiratory and Sleep Medicine, Mater Misericordiae Children's Hospital of South Brisbane in the state of Queensland of Australia.

REFERENCE

[1] Anch, AM; Browman, CP; Mitler, MM; Walsh, JK. Sleep: a scientific perspective. *Englewood Cliffs, NJ: Prentice-Hall*; 1988.

[2] Hobson, JA. Sleep. New York, NY: *Scientific American Library,* 1989

[3] Hobson, JA. Sleep is of the brain, by the brain and for the brain. *Nature,* 2005, 437, 1254-1256.

[4] Deboer, T. Technologies of sleep research. *Cell Mol Life Sci*, 2007, 64, 1227-1235.

[5] Hill, CM; Hogan, AM; Karmiloff-Smith, A. To sleep, perchance to enrich learning? *Arch Dis Child*, 2007, 92, 637-643.

[6] Empson, J. Sleep and dreaming, 3rd Edition. Houndmills, Basingstoke, Hampshire, England: *Palgrave Publishers*, 2002.

[7] Fauroux, B; Aubertin, G; Clement, A. What's new in paediatric sleep in 2007? *Paediatr Respir Rev*, 2008, 9, 139-143.

[8] Gozal, D. Obstructive sleep apnea in children: implications for the developing central nervous system. *Semin Pediatr Neurol*, 2008, 15, 100-106.

[9] Meltzer, LJ; Mindell, JA. Sleep and sleep disorders in children and adolescents. *Psychiatr Clin North Am*, 2006, 29, 1059-1076.

[10] Central Intelligent Agency. *The world factbook*. Washington, DC; 2008

[11] Espenshade, TJ; Guzman, JC; Westoff, CF. The surprising global variation in replacement fertility. *Popul Res Policy Rev*, 2003, 22, 575-583.

[12] Vander, AJ; Sherman, JH Luciano, DS. Human physiology: the mechanisms of body function, 8th Edition. Columbus, OH: *McGraw-Hill*, 2001.
[13] Marieb, EN. Human anatomy and physiology, 6th Edition. San Francisco, CA: *Pearson Benjamin Cummings*, 2004.
[14] Martini, F. Fundamentals of anatomy and physiology, 5th Edition. Upper Saddle River, NJ: *Prentice Hall*, 2001.
[15] McNamara, F; Sullivan, CE. Evolution of sleep-disordered breathing and sleep in infants. *J Paediatr Child Health*, 1998, 34, 37-43.
[16] Loughlin, GM. Obstructive sleep apnea in children. Adv Pediatr, 1992, 39, 307-336.
[17] Pappenheimer, J. Handbook of physiology: a critical, comprehensive presentation of physiological knowledge and concepts, section 3, volume 2. *Bethesda, MD*: *American Physiological Society*, 1977.
[18] Hudgel, DW. Mechanisms of obstructive sleep apnea. *Chest*, 1992, 101, 541-549.
[19] Ayappa, I; Rapoport, DM. The upper airway in sleep: physiology of the pharynx. *Sleep Med Rev*, 2003, 7, 9-33
[20] Victor, LD. Obstructive sleep apnea. *Am Fam Physician*, 1999, 60, 2279-2286.
[21] Kimoff, RJ. Sleep fragmentation in obstructive sleep apnea. *Sleep*, 1996, 19, S61-S66.
[22] Mendelson, WB. Human sleep: research and clinical care. New York, NY: Plenum Medical Book; 1987.
[23] Bradley, TD; Floras, JS. Sleep apnea and heart failure: Part II: *Central sleep apnea, Circulation*, 2003, 107, 1822-1826.
[24] Bradley, TD. Crossing the threshold: Implications for central sleep apnea. *Am J Respir Crit Care Med*, 2002, 165, 1203-1204.
[25] Bradley, TD; Phillipson, EA. *Central sleep apnea, Clin Chest Med*, 1992, 13, 493-505.
[26] Kryger, MH. Central apnea. *Arch Intern Med*, 1982, 142, 1793-1794
[27] Nitzan, M; Khanokh, B; Slovik, Y. The difference in pulse transit time to the toe and finger measured by photoplethysmography. *Physiol Meas*, 2002, 23, 85-93.
[28] Smith, RP; Argod, J; Pepin, JL; Levy, PA. Pulse transit time: an appraisal of potential clinical applications. *Thorax*, 1999, 54, 452-458
[29] Kahn, A; Groswasser, J; Sottiaux, M; Rebuffat, E; Franco, P. Mechanisms of obstructive sleep apneas in infants. *Biol Neonate*, 1994, 65, 235-239.
[30] Gozal, D. Obstructive sleep apnea in children, *Minerva Pediatr*, 2000, 52, 629-639.

[31] Stocks, J; Sly, PD; Tepper RS; Morgan WJ. Infant respiratory function testing. Wilmington, DE: *Wiley-Liss Inc*, 1996.

[32] Rosen, CL. Obstructive sleep apnea syndrome (OSAS) in children: Diagnostic challenges. *Sleep*, 1996, 19, S274-S277.

[33] Schechter, MS; Section on Pediatric Pulmonology, Subcommittee on Obstructive Sleep Apnea Syndrome. Technical report: diagnosis and management of childhood obstructive sleep apnea syndrome. *Pediatrics*, 2002, 109, e69.

[34] Stores, G; Wiggs, L. Sleep disturbance in children and adolescents with disorders of development: its significance and management. *Cambridge, UK: Mac Keith Press*, 2001.

[35] Blunden, S; Lushington, K; Kennedy, D; Martin, J; Dawson, D. Behavior and neurocognitive performance in children aged 5-10 years who snore compared to controls. *J Clin Exp Neuropsychol*, 2000, 22, 554-568.

[36] Meijer, AM; Habekothe, HT; Van-Den-Wittenboer, GLH. Time in bed, quality of sleep and school functioning of children. *J Sleep Res*, 2000, 9, 145-153.

[37] Richards, W; Ferdman, RM. Prolonged morbidity due to delays in the diagnosis and treatment of obstructive sleep apnea in children. *Clin Pediatr*, 2000, 39, 103-108.

[38] Pitson, D; Chhina, N; Knijn, M; Herwaaden, V; Stradling, J. Changes in pulse transit time and pulse rate as markers of arousal from sleep in normal subjects. *Clin Sci*, 1994, 87, 269-273.

[39] Manser, RL; Rochford, P; Naughton, MT; Picrce, RJ; Sasse, A; Teichtahl, H; Ho, M; Campbell, DA. Measurement variability in sleep disorders medicine: the Victorian experience. *Intern Med J*, 2002, 32, 386-393.

[40] Kapur, V; Blough, DK; Sandblom, RE; Hert, R; De Maine, JB; Sullivan, SD; Psaty, BM. The medical cost of undiagnosed sleep apnea. Sleep 1999, 22, 749-755.

[41] Rechstschaffen, A; Kales, A. A Manual of Standardized Terminology, Techniques and Scoring Systems for Sleep Stages of Human Subjects. Washington, DC: Public Health Service, Government Printing Office, 1968.

[42] American Thoracic Society, Medical Section of the American Lung Association. Standards and indications for cardiopulmonary sleep studies in children. *Am J Respir Crit Care Med*, 1996, 153, 866-878.

[43] AASM Clinical Practice Review Committee. *Cost justification for diagnosis and treatment of obstructive sleep apnea, Sleep*, 2000, 23, 1017-1018.

[44] Escourrou, P; Luriau, S; Rehel, M; Nedelcoux, H; Lanoe, JL. Needs and costs of sleep monitoring. *Stud Health Technol Inform*, 2000, 78, 69-85

[45] Man, GC; Kang, BV. Validation of a portable sleep apnea monitoring device. *Chest*, 1995, 108, 388-393.
[46] Teschler, H; Berthon-Jones, M. Full polysomnography versus home sleep study: Searching for the optimal procedure. *Eur Respir J*, 1997, 10, 1699-1700.
[47] Chervin, RD; Murman, DL; Malow, BA; Totten, V. Cost-utility of three approaches to the diagnosis of sleep apnea: polysomnography, home testing, and empirical therapy. *Ann Intern Med*, 1999, 130, 496-505.
[48] Mykytyn, IJ; Sajkov, D; Neill, AM; McEvoy, RD. Portable computerized polysomnography in attended and unattended settings. *Chest*, 1999, 115, 114-122.
[49] Fry, JM; DiPhillipo, MA; Curran, K; Goldberg R; Baran AS. Full polysomnography in the home. *Sleep*, 1998, 21, 635-642.
[50] Ajilore, O; Stickgold, R; Rittenhouse, CD; Hobson, JA. Nightcap: Laboratory and home-based evaluation of a portable sleep monitor. *Psychophysiology*, 1995, 32, 92-98.
[51] Lloberes, P; Sampol, G; Levy, G; Aristizabal, D; Sagales, T; De la Calzada, M; Roca, A; Canas, I; Morell, F. Influence of setting on unattended respiratory monitoring in the sleep apnoea/hypopnoea syndrome. *Eur Respir J*, 2001, 18, 530-534.
[52] Standards of Practice Committee of the American Sleep Disorders Association. Practice parameters for the use of portable recording in the assessment of obstructive sleep apnea. *Sleep*, 1994, 17, 372-377.
[53] White, DP. Complex home monitoring. *Sleep*, 1996, 19, S248-S250.
[54] Golpe, R; Jimenez, A; Carpizo, R; Cifrian, JM. Utility of home oximetry as a screening test for patients with moderate to severe symptoms of obstructive sleep apnea. *Sleep*, 1999, 22, 932-937.
[55] Baltzan, MA; Verschelden, P; Al-Jahdali, H; Olha, AE; Kimoff, RJ. Accuracy of oximetry with thermistor (Oxiflow) for diagnosis of obstructive sleep apnea and hypopnea. *Sleep*, 2000, 23, 61-69.

[56] Lorenzo, JL; Barbanoj, MJ. Variability of sleep parameters across multiple laboratory sessions in healthy young subjects: the "very first night effect". *Psychophysiology*, 2002, 39, 409-413.
[57] Scholle, S; Scholle, HC; Kemper, A; Glaser, S; Rieger, B; Kemper, G; Zwacka, G. First night effect in children and adolescents undergoing polysomnography for sleep-disordered breathing. *Clin Neurophysiol*, 2003, 114, 2138-2145.
[58] Saletu, B; Klosch, G; Gruber, G; Anderer, P; Udomratn, P; Frey, R. First-

night-effects on generalized anxiety disorder (GAD)-based insomnia: laboratory versus home sleep recordings. *Sleep*, 1996, 19, 691-697.

[59] Toussaint, M; Luthringer, R; Schaltenbrand, N; Carelli, G; Lainey, E; Jacqmin, A; Muzet, A; Macher, JP. First-night effect in normal subjects and psychiatric inpatients. *Sleep*, 1995, 18, 463-469.

In: Biological Clocks: Effects on Behavior... ISBN: 978-1-60741-251-9
Editors: O. Salvenmoser et al. pp. 101-120

Chapter 4

CIRCADIAN SLEEP-WAKE RHYTHMS IN PRETERM INFANTS

Ronny Geva and Ruth Feldman
The Gonda Multidisciplinary Brain Research Center
Bar-Ilan University, Israel

ABSTRACT

Knowledge is recently building regarding the factors that affect the development of an optimal sleep-wake cycle. Fetuses near term experience arousal fluctuations and sleep episodes already in the darkness of the womb. Yet, the factors that allow for a smooth emergence of an optimal sleep wake cycle are the conditions that place newborn at higher risk to develop sleep disorders are not yet fully understood. The current chapter proposes an integrative model for the development of circadian sleep-wake rhythms in preterm infants. This model is based on a comprehensive review of the basic science literature as well as clinical work with at-risk preterm infants. This integrative model proposed a three source risk source: 1) infant dependent neurobiological vulnerability risk, such as brainstem mediated functional disability; 2) a familial source that entails genetic predispositions, parental style and support resources; and 3) an environmental stimulation source, such as excessive NICU related stimulation and handling. These three realms act as main effects and interact with the infants' prenatal and postnatal age and with the infants' self- regulation mechanisms. The manner in which these sources operate and interact lead to directions of interventions that ameliorate the development of circadian sleep-wake rhythm deficit in infants born at-risk for sleep disorders. The chapter details each component of the model and illuminates the complexity of the interactions involved in it to deepen the understanding the mechanisms involved in early organization of sleep-wake rhythms in infants.

INTRODUCTION

Sleep-wake rhythms serve as a basis for the modulation of arousal (Feldman et al., 2002; Feldman et al., 2004), which in turn provides the foundation for emotion regulation (Kopp, 1999; Feldman, Weller, Sirota, & Eidelman, 2002), social competencies (Feldman, 2006) attention shifting and maintenance (Aston-Jones, 2005), sustained exploration (Feldman et al., 2002) and cognition (Sadeh, Dark, Vohr, 1996; & Mayes, 1999; Scher, Amir & Tirosh, 2000), by providing a global organizing framework of arousal. The development of circadian sleep – wake rhythms is therefore of central importance in securing adaptive development across infancy.

Noradrenergic locus coeruleus (LC) neurons provide a circadian regulation of the sleep-wake cycle, and the maintenance of LC function depends on light exposure. Studies have shown that noradrenergic LC neurons exhibit a circadian rhythm in impulse activity, which peaks during the active period. This is mediated by an indirect circuit projection from the suprachiasmatic nucleus (SCN) to the LC, via the dorsomedial and paraventricular hypothalamic nuclei, as well as medial and ventrolateral pre-optic areas (Aston-Jones, Chen, Zhu, & Oshinsky, 2001).

Light deprivation induces a loss of noradrenergic fibers, which in turn decreases the amplitude of the sleep-wake rhythm (Gonzalez & Aston-Jones, 2006). The effects of excessive light exposure on sleep wake rhythmicity were recently described in a case study (Doljansky, Kannety & Dagan, 2005). Modifications of this exposure were shown to yield normalization effect of the sleep wake cycle within days. Effects of excessive light exposure, particularly during night hours on the neonatal circadian sleep-wake cycle are presently debated.

FETAL ENTRAINMENT

In recent years, experimental data shows that among the significant developmental changes occurring in the 2nd half of pregnancy, sleep related structural and functional developments are among the most central. Recent findings present intriguing findings that fetal endogenous system is capable of generating increasingly differentiated fetal behavioral states and transitions between them in association with fetal development, such clearly defined non-REM and REM conditions (Frank et al., 2006). These differential bio-behavioral states are increasingly evident by short epochs of heart rate acceleration and deceleration that define differential regularity patterns (Kuhnert, Hellmeyer, Stein, Schmidt, 2006); Van Leeuwen, Cysarz, Lange & Gronemeyer, 2006)

Halpern and colleagues in their review (Halpern, Maclean and Baumeister ,1995) noted that the chronological age at which binary sleep – wake states first appear are reported to occur at different phases, depending upon methodological differences. Studies conducted in the 70s noted that active sleep emerges at about 32-35 weeks gestation (Dreyfus-Brisac, 1970;Parmelee, Wenner, Akiyama, Schultz & Stern, 1967) and quiet sleep emerges a bit later at about 32-35 weeks of gestation (Dreyfus-Brisac, 1970; Prechtl, 1974). However, studies conducted in the 90s showed that both active and quite sleep can be reliably detected in both healthy and medically vulnerable preterm infants that are 28-31 weeks old (Curzi-Dascalova et al., 1988; Holditch-Davis, 1990; Stefanski, Schulze, Bateman, Kairam, Pedley, Masterson, & James, 1984).

Yet more recent studies from the last 5 years indicate that cyclicity of neonatal sleep behaviors emerges for most at 25 to 30 weeks' post-conceptional age, reflecting an ultradian biologic rhythm during the early perinatal stage of brain development (Scher, Johnson, Holditch-Davis, 2005).

It has been recently shown that the fetal sleep-wake system is also capable of producing in vivo intrauterine circadian rhythms during the third trimester (Frank et al., 2006). This development is to some degree endogenously driven (Mirmiran, Baldwin, Ariagno, 2003). It is also operating via responding to maternal entrainment signals of day-night rhythms. This circadian rhythmicy is measurable via recordings of fetal heart rate fluctuations in human fetuses for various periods of times. This capacity is not an expression of a free running internal clock, despite the relative darkness, typical of uterine environment. It represents effects related to fetus' capability of sensing the maternal diurnal rhythm, through reception of rhythmic auditory changes, bowel movement changes, HR changes, and more. Analysis of the results is indicative of some preliminary rhythmicity.

In primates, maternal melatonin has been suggested as a candidate to entrain fetal circadian rhythms, including the SCN rhythms of metabolic activity. Recently maternal melatonin has been indeed shown to be a Zeitgeber for the fetal supra charismatic nucleus of the hypothalamus, but probably not for the adrenal gland (Torres-Farfan, Rocco, Monso, Valenzuela, Campino, Germain, et al., 2006).

Neonatal Entrainment

McGrow and colleagues (1999), following the emergence of entrainment of a healthy male infant in Texas from birth to 6 months of age in conditions of a household with a fixed schedule and natural light environment showed that the circadian rhythm of temperature appeared first, during the first week of life. The wake circadian rhythm appeared second, attaining significance at day 45; approximately at the same time that increased melatonin concentration began to occur at sunset. The sleep circadian rhythm appeared last, attaining significance after day 56. Ninety to 120 minute zones of sustained wakefulness first appeared in the second month of life subsequent to awakening and prior to sleep onset. The infant's nocturnal sleep-onset was coupled to sunset before day 60 and subsequently to family bedtime, giving evidence of initial photic entrainment followed by social entrainment. McGraw and colleagues concluded that Circadian rhythms appeared much more rapidly in this infant than previously reported; their rapid appearance was probably facilitated by maximal exposure to sunlight, and presence of social routine cues, such as schedules of family meals, bath time and such, that were closely tied to daylight cyclicity. Thus McGraw's work underscores the importance of exposure to artificial light and irregular handling schedules as risk factors that delay normal circadian rhythm attainment (McGraw, Hoffmann, Harker & Herman, 1999).

An Integrative Model for the Development of Circadian Sleep-Wake Rhythms in Preterm Infants

Based on reviewed literature that will be discussed in detail below, we hypothesize an integrative model that integrates neurobiological vulnerability, maturation and stimulation factors that eventually pattern circadian sleep-wake rhythms in neonates. This model is based on an extensive literature review of both basic science models as well as on clinical studies with normally developing and at-risk human neonates.

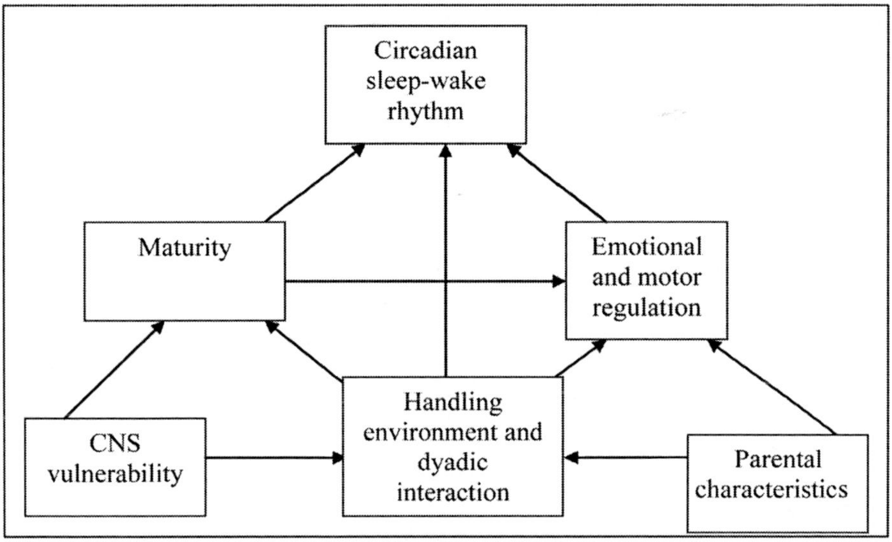

Figure 1. An integrative model for the development of circadian sleep-wake rhythms in preterm infants

This integrative model proposed a three source risk source: 1) infant dependent neurobiological vulnerability risk, such as brainstem mediated functional disability; 2) a familial source that entails genetic predispositions, parental style and support resources; and 3) an environmental stimulation source, such as excessive NICU related stimulation and handling. These three realms act as main effects and interact with the infants' prenatal and postnatal age and with the infants' self- regulation mechanisms. The later then mediate the development of infants' circadian sleep-wake rhythm. Further understanding of the manner in which these sources operate and interact would lead to development of proper interventions to eliminate the development of circadian sleep-wake rhythm deficit in infants born at-risk for sleep disorders. In the following we detail each component of the model

Prematurity - A Unique Window to Study Preliminary Circadian Rhythm Development

Maternal illness and premature birth adversely affect the development of circadian rhythms (Mirmiran & Ariagno, 2000). These effects seem to be strongest around the developmental mile stone at which the adverse process, .i.e., preterm birth, occurred (Mirmiran & Ariagno, 2000). The authors note this process disrupts the initial mother-infant process which is also vital for proper entrainment.

Prematurely born infants offer a unique window to study the circadian rhythm related developmental progressions. However, studies on sleep patterns of preterm infants at the age of expected birth have shown a fairly variable pattern of results (Ingersoll & Thoman, 1999). For example, on some patterns, preterm neonatal measures appeared to be precaucial relatively to neonates born at term, while on other parameters, an opposite direction was apparent (Booth Thoman & Leonard 1980; Holditch-Davis & Thoman, 1987). So that around birth, preterm neonates showed a higher frequency of REM in active sleep, and their sleep was characterized by longer quite sleep episodes.

It is worth noting that some similarities in circadian behaviors occurred in the development of sleep-wake cycle in preterm infants in the extra-uterine environment and full-term infants in the intrauterine environment (de Vries, Visser, & Prechtl, 1982; Diambra & Menna-Barreto, 2004; Olischar, Klebermass, Kuhle, Hulek, Kohlhauser, Rucklinger, Pollak & Weninger, 2004). However, recent data point to a vital transitional milestone that occurs around 31 weeks of gestation. This milestone, points to a possible additional risk related to preterm birth prior to 31 weeks of gestation. Following the development of sleep-wake cyclicity in premature infants born at 25 to 29 weeks gestation, we found that a developmental leap in the organization of the sleep-wake cycle occurred at 31 weeks gestational age and that those infants who did not show a timely maturation of the biological clock showed deficiencies in social interactions and arousal modulation capacities in later infancy (Feldman, 2006).

Overall, as noted in Halpern, Maclean & Baumeister's review (Halpern et al., 2006) the variability in sleep parameters in preterm infants is much greater than that reported in infants born at term. This tendency holds also in preterm infants born extremely early and or with extremely small weights (Leitner, Bloch, Sadeh, Neuderfer,Tikotzky, Fattal-Valevski et al., 2002). Researchers thus continue to search for additional risk factors that would explain significant portions of this large variability. Among the variables recently explored both infants factors and environmental measures are studied.

CNS Perinatal Susceptibility as a Risk Factor

Brainstem structures regulate and modulate sleep and arousal (Losier & Semba, 1993). Structures in the brainstem influence higher brain structures in the midbrain and cortex. The ascending reticular activating system that transverses through the brainstem stimulation to the thalamus, hypothalamus, and basal forebrain, results in excitation of the cortex. Its cholinergic, glutamic and catacholinergic neurons are stimulated by afferent sensory input, such as audition and vision (Harris, 2005). Wakefulness is manifested by an active cortex that exhibits a characteristic pattern of desynchronized EEG

Mechanisms related to the nucleus tractus solitarius, which project on to midbrain and forebrain structures, inhibit activity in the reticular system, resulting in activation of inhibitory thalamocortical projections to the cortex. During a state of decreased activation, the cortex exhibits a pattern of synchronized EEG that is manifested as various forms of Non-REM and REM sleep (Harris, 2005).

Other brainstem structures regulate activities known to affect sleep and arousal indirectly, such as arousal and attention (Lai, Shalita, Hajnik, Wu, Kuo, Chia, et al., 1999; Wong & Wong, 1991; McNamara, Wulbrand & Thach, 1999; Karmel, Gardner & Freedland, 1996) and the processing of sensory information (Anthony & Graham, 1997; Nattie, 2000), such as visual and auditory perception (Larson & Yajima, 1994; Jiang, & Wilkinson, 2006; Aitkin, 1986; Davidson & Bender, 1991). Still, direct effects of brainstem dysfunction on sleep and arousal regulation early on in development are not yet fully understood.

Neonatal brainstem dysfunction has been suggested as a risk factor for various neural-developmental deficits (Karmel, Gardner, Zapulla, Magnano & Brown, 1988; Darnall, Ariagno, Kinney, 2006): neurobehavioral deficits (Karmel, Gardner, Zappulla, Magnano, & Brown, 1988), orodigestive and cardiorespiratory difficulties (Abadie, Morisseau-Durand, Beyler, Manach & Couly, 2002; Darnall, Ariagno, Kinney, 2006), and neurological dysfunction (Yilmaz, Degirmenci, Akdas, Kulekci, Ciprut, Yuksel, et al., 2001). We propose that brainstem dysfunction may be directly related to an aberrant development of sleep- wake cycles, particularly in preterm neonates for whom the maturation of higher-order structures is still under way. For such infants, the sensitive circadian system may have been compromised by the pre-term birth.

It has been shown that Cholinergic and aminergic brainstem neurons can concurrently modulate the activity of neurons in the thalamus and basal forebrain during cortical arousal, through dual projections (Losier & Semba, 1993- in the rat model). Structural damage in brainstem structures, specifically in the ventral mesencephalon and rostroventral Pons have been shown to results in deficits in sleep- wake organization (Lai, Shalita, Hajnik, Wu, Kuo, Chia & Siegel, 1999- in the cat model). It is thus hypothesized that brainstem-related neonatal dysfunctions would result in abnormalities in the development of the sleep-wake circadian rhythms.

NICU Environment as a Risk Factor

The environment of the NICU presents ample demands on the preterm infants (Field, 1990). The NICU's great life-saving potential comes often at some cost. The NICU environment is different from maternal uterus in numerous fashions, some of particular importance to circadian rhythm development.

The NICU presents various levels and types of lights that are significantly different from natural lights. Solar lights are presented un-baffled by uterine wall and artificial light, at various wave lengths, are often intense and are presented at prolonged phases, way beyond characteristics of solar light phase (Niessen, 2006).

The NICU environment also presents the infant with excessive noise (Kent, Tan, Clarke & Bardell, 2002; Byers, Waugh & Lowman, 2006; Niessen, 2006). It is fraught with multiple noises, monitor beeping, ventilator alarms, low spectrum noises of wheeling incubators and various carts during the day and the night and stuff activity related noises. Some are fairly constant through the day and the night, some are impulsive and unexpected. Overall activity levels are higher than other environments to which infants are brought to. Still, recent measurements of sound pressure levels at NICU indicated the highest levels were recorder at 6-7 AM and around noon (Kruege, Wall, Parker, & Nealis, 2005). Activity at NICU is maintained at relatively high levels during the day and also is significantly high during the night, though typically lower than that of the day.

The effects of loud noise on the preterm infants are evident on a physiological level by increased heart and respiratory rates and decreased oxygen saturation (Bremmer, Byers, & Kiehl, 2003). Its effects on sleep –wake rhythms are not yet fully understood. However, animal models have demonstrated via simulating continuous lights and sounds of a typical nursery and exposing newborn animals to these conditions, showed that this constellation of stimuli schedule results in permanent damage to the biological clock, in differences in sensory processing of stimuli, and in immature habituation to stimuli in the young animal (Hao & Rivkees, 1999- in a baboon's model; Philbin, Ballweg, & Gray, 1994 and Sleigh & Lickliter, 1998- in chicks models). The studies point to the potentially damaging effects of nursery conditions on the developing infant in general and have a potential harmful effect on its developing systems that regulate circadian activity.

Experimental light manipulation has shown, that the circadian sleep-wake rhythm in the preterm infants entrained after a similar time of exposure to an environment with daily time cues but at an earlier post-conceptional age when compared with the term group (McMillen, Kok, Adamson, Deayton & Nowak, 1991). While McMillen's group in Australia has concluded that maturity of the premature infant is less of a determinant in attaining circadian sleep, but rather that the length of time spent in a cycles day-night environment is vital for this purpose (McMillen, Kok, Adamson, Deayton & Nowak, 1991).

Environmental variability at the NICU is often less directly related to the status of the infants, but rather to necessary management of neighboring infants and the fixed routine procedures of the NICU, irrespective of the infants' current status. This factor is also very different from uterine environment where by the uterine environment is almost solely responsive to the mother-infant dyad capacity, changes and needs. It may very well be that a chaotic, non-circadian environment of the NICU adversely affects initiation and stabilization of circadian rhythms of the preterm infants (Mirmiran & Ariagno, 2000).

Attempts to support this hypothesis were so far only partly successful, since circadian and sleep development in preterm infants occurs independently from the influences of environmental lighting (Mirmiran, et al., 2003). Entrainment of preterm infants using a different lighting environment during the night (a day-night cycled room), seem to be not effective or at least not sufficient to differentially effect entrainment during the first 3 months of the preterm infants post natal life (Mirmiran, et al., 2003). Similarly, phototherapy at the NICU has been shown to have no short-term effects, on neonatal sleep-wake cycles when compared with preterm infants who have not received photic intervention. Nor does it effect sleep-wake rhythms or saliva cortisol levels at two years of age (Shimada, Segawa, Higurashi, Kimura, Oku, Yamanami et al., 2003).

At the same time, analysis of activity levels recorded by Actiwatches, cycled light (7PM-7AM use of crib covers) has been shown to affect activity levels during the first month post preterm birth (Rivkees, 2003; Rivkees, Mayes, Jacobs & Gross, 2004). Exposure of premature infants to low-intensity cycled lighting, rather than to a fixed dim light level, in the hospital nursery induces distinct patterns of rest-activity that are apparent within 1 week after discharge. The data thus shows that day-night rhythms in activity patterns can be detected shortly after discharge to home in premature infants and that the circadian clock of developing infants is entrained by cycled lighting.

Harris (2005) in his review summarizes that the amount of time spent sleep changes throughout the life span. Such, that at the neonatal-period the infant sleep most of time. Sleep periods of 1-4 hours are distributed evenly during the circadian cycle. By 6 months of age well defined sleep-activity states appear. Sleep episodes lengthen and the proportion of REM sleep declines (Harris, 2005)

Birth Mode as a Risk Factor

Effects of specific risk factors, which typically accompany preterm birth on circadian sleep wake rhythms is yet significantly understudied. Of the various significant risk factors for which some preliminary data has been collected, one can find: Delivery mode, that seem to be of no consequence on the amount of sleep but rather has an effect on the development of circadian rhythms in neonates (Korte, Hoehn & Siegmund, 2004). It was found that the majority of vaginally born neonates showed a distinct circadian frequency in their spectra, which is characterized by more sleep during the night. In contrast, both groups of neonates born by C-section showed significantly less distinct circadian frequencies in their spectra. The authors did not find these differences in infants born preterm.

Others find that c-section deliveries are related to differences in sleep- wake organization during the first few days post birth.

In Finland, postnatal adaptation after c-section was found to be related to changes in sleep organization, and increases in oxyhaemoglobin saturation and frequency of body movements in the first few days after birth (Nikkol, Kirjavainen, Ekblad, Kero & Salonen, 2002). Similar findings were reported in the US, where by full term neonates born via vaginal delivery had more wakefulness, shorter mean sleep periods and shorter longest-sleep periods during the daytime on the first two days post birth (Freudigman & Thoman, 1998) . Thus, perinatal interventions such as stress and pain (Nikkol, Kirjavainen, Ekblad, Kero & Salonen, 2002) and surgical birth (Freudigman & Thoman, 1998) seem to disrupt neonatal circadian rhythm of full term neonates. The mechanism/s of these effects is not yet understood.

Prematurity as a Risk Factor

Effects of prematurity of sleep-wake rhythms seem to be complex. On the one hand one may hypothesize greater vulnerability of a younger system in handling excessive interventions. At the same time greater plasticity to aberrations is to be expected under certain conditions. Generally, preterm infants exhibit more active sleep and indeterminate sleep states than full-term infants (Curzi-Dascalova, Peirano, & Morel-Kahn, 1988; High & Gorski, 1985; Holditch-Davis, 1990; Holditch-Davis & Edwards, 1998; Parmelee, Wenner, Akiyama, Schultz, & Stern, 1967). At the same time, there is a significant debate regarding indications of sleep-wake rhythmicity in preterm neonates.

Following these robust findings, some presume a greater weight for prematurity than to its related complications, such as illness severity and medical treatments, handling and hospitalization (Holditch-Davis, Scher, Schwartz, Hudson-Barr, 2004). Indeed, large scale studies, such as that performed by Sadeh, show that sleep of infants born at later gestational ages is characterized of greater organization levels, indicated by increased percent of quiet and motionless sleep (Sadeh, et al., 1996). Yet, it is not know whether the effect of prematurity on sleep-wake rhythms is a main effect one, or rather that risk factors interact with gestational age to affect sleep-wake cycles during the neonatal period. An example for such an interaction effect is maternal glucose values during pregnancy (Sadeh, et al., 1996) - which is related to sleep-wake patterns in preterm infants, but not in the controls. Another factor recently discovered that interacts with prematurity to affect sleep characteristics in preterm infants is administration of antenatal magnesium sulfate and corticosteroid therapy (Black, Holditch-Davis, Schwartz & Scher, 2006). It was found infants exposed to magnesium sulfate had more active sleep without rapid eye movement, indicating poorly organized active sleep. Investigators seem to concur that the mechanism involved in the majority of these processes is some subtle affect on CNS in these vulnerable infants.

The Development of Circadian Sleep-Wake Rhythm in Preterm Infants as a Factor of Parenting and Environmental Stimulation

Postnatal entrainment goes through marked changes during the first few weeks of life. Entrainment to maternal schedule occurs irrespective of prematurity around the post conceptual age of 45 weeks in 75% of infants (Shimada, Takahashi, Segawa, Higurashi, Samejim & Horiuchi, 1999). During the transition periods infants go through either ultradian of irregular sleep-wake cycles (Shimada, Takahashi, Segawa, Higurashi, Samejim & Horiuchi, 1999).

Neonatal cyclicity measures are reported to be negatively related to indices of advanced perinatal status as well as 6-month mental scores; however, at 6 months, the cyclicity measures from infant born preterm were positively related to perinatal and neonatal measures as well as mental scores (Borghese, Minard & Thoman, 1995; Feldman, 2006). These seemingly contradicting results may indicate the necessity for different interpretations of periodicity at the preterm and later age (Borghese, Minard & Thoman, 1995).

Sleep disorders have been reported in children born preterm as they grow up. A study conducted at 20 months corrected age has shown that the quality of nighttime sleep of preterm toddlers is more disrupted on average than that of toddlers born at term (Gossel-Symank, Grimmer, Korte, Siegmund, 2004). The daytime rest duration of preterm infants at 20 months is significantly shorter than that of full-term infants. The sleep quality of preterm infants is significantly lower than that of full-term infants, since the preterm infants have a larger percentage of less restful nighttime sleep (Gossel-Symank, Grimmer, Korte, & Siegmund, 2004).

A theoretical framework that may account for the complex circadian sleep-wake rhythm of neonates proposes an interaction between components that are more directly related to CNS vulnerability and other components that are more prone to handling and stimulation effects (Anders, Keener, Kraemer, 1985)

Literature cited by Halpern (Halpern et al., 1995) shows that during the 90s it was thought that the first 6 months of life are not yet accompanied by changes in the awaked state during nighttime (Anders & Keener, 1985). Halpern summarizes reports of observational studies of infants during the day time during the first 6 months of postnatal life, in which a non-linear increase in the percent of day time wakefulness, and longer periods of wakefulness during the day (Halpern, 1995; Wolff, 1984). Comparisons of wakefulness at daycare settings and in the home revealed that wakefulness in the home setting was maintained at earlier ages and for longer periods of times (Wolff, 1984) This difference was attributed to the level of stimulation provided in the home relatively to relatively lack of it at the day care setting (Thoman & Whitney, 1989). Stimulation effect on wakefulness was also found as a factor of maternal company, parental co-sleeping (Thoman, 2006); as opposed to a lonely setting.

The age of early marks of circadian influences on sleep-wake patterns is still in debate (Halpern, 1995). Some reported on first appearance between 6-8 weeks of age (Thoman & Whitney, 1989), even though gradual increase in daylight wakefulness emerges earlier, at around 4-5 weeks of age in the home setting (Thoman & Whitney, 1989). Significant sleep episodes during the night are typically fairly common at 3 months of age (Coons & Giullenault, 1984).

Effects of Touch Interventions for Preterm Infants on Sleep-Wake Circadian Rhythmicity

The development of sleeping and waking behaviors in preterm infants is reactive to handing and to social encounters right from the neonatal period (Ingersoll & Thoman, 1999). Nursing care affects sleep states already at the nursery and during the first three months of post natal life (Brandon, Holditch-Davis & Beylea, 1999).

Of special interest are the effects of touch-and-contact interventions on sleep-wake rhythmicity in neonates.

Among the central benefits of touch-and-contact intervention to the fragile premature infant is the organization of the biological clock. Animal studies point to the important role of the mother's physical proximity and specific body cues for the maturation of the biological clock in the pup during the first post-birth days (Hofer, 1995). In addition to their immature CNS, premature infants are deprived of the soothing, regulating, and organizing presence of the mother's body and closeness (Feldman, 2004). Two types of touch-and-contact interventions were found to have an impact on sleep-state organization; Kangaroo Care and massage.

Kangaroo Care (KC), a method developed in Bogota, Columbia during a period of incubator shortage, involves skin-to-skin contact between mother and her premature infant. Several studies showed both concurrent and long-lasting effects to the KC intervention on state organization and sleep-wake cyclicity. During contact, infants spent more time in quiet sleep, heart rate was lower and more stable, apnea and bradycardia decreased, body temperature was maintained, and oxygenation improved. Moreover, the effects of skin-to-skin contact seemed to persist even after contact was ended. Following kangaroo contact premature infants slept longer and sleep was more restful and organized. Interestingly, KC improved not only sleep states but also infant alertness, as shown by the findings that infants spent longer periods in alert states after KC. It is thus possible that KC organizes the biological clock by increasing the time infants spend in the two ends of the arousal continuum—quiet sleep and alert wakefulness—while reducing transitory states and active sleep. An addition contribution of kangaroo contact to state organization was the decrease in transitory, indeterminate sleep states among infants born preterm who underwent KC as compared to matched controls (Feldman & Eidelman, 2003).

Kangaroo care similarly contributed to the organization of sleep-wake rhythmicity. Following two weeks of daily kangaroo contact, we found that the amplitudes of sleep-wake cyclicity as measured in 10-second epochs over four consecutive hours were higher for treated infants compared to controls (Feldman et al., 2002). Better organization of the biological clock, in turn, was related to improved arousal and emotion regulation capacities across infancy and up to age five. Infants with better sleep-wake organization had higher thresholds to negative stimuli, were better able to modulate arousal, and showed more exploratory behavior across the first year. During the toddler and preschool age, these children scored higher on neuropsychological tests that required attention shifting and maintenance and displayed higher executive functioning at age five. In the social-emotional domains, children with better organization of the biological clock were more able to show restrained and self-regulation during social and nonsocial tasks that requires inhibition and regulation of anger and frustration (Feldman, 2007).

Massage therapy, active stroking of the premature infants following a specific regime similarly showed that premature infants who received massage therapy spent more time in alert wakefulness and less time in active sleep states, were discharged earlier from the hospital, and showed more mature neurobehavioral profiles (Field, 1995).

Overall, recent developments demonstrate intervention effects of stimulation, on sleep-wake state organization and the development of sleep-wake circadian rhythms of infants born prematurely, via careful modifications of infant's exposure to light, sound, and touch in early phases of development. Developing multi-modality intervention protocols that are more closely tied to circadian mechanisms of the infant and its CNS capacities may re-direct development of infants born at risk into more adaptive sleep-wake circadian rhythms.

CONCLUSION

We reviewed and critically tested the primary issues affecting the development of sleep-wake circadian rhythms in infants: 1) CNS vulnerability, as indexed by specific brain structures known to be involved in regulation of sleep-wake functions, such as *perinatal brainstem dysfunction*; 2) *Aberrant maturation caused by premature birth*; and, 3) *Suboptimal environmental conditions*. This topic relates to the unnatural neonatal environment and excessive exposure to light sound and touch in early development.

We propose that an integrative view that takes into account the vulnerability of autonomic system, neonatal neural maturation, and environmental conditions provide the foundation for the emergence of circadian sleep patterns in early development. Finally, this source has been proposed as a valuable intervention tool, in the forms of specific message and KC in ameliorating the adverse effects of CNS related vulnerability, and excessive stimulation early on in development and redirecting a more optimal development of neonatal circadian sleep-wake cycles.

References

Abadie, V., Morisseau-Durand, M. P., Beyler, C., Manach, Y. & Couly, G. (2002). Brainstem dysfunction: a possible neuroembryological pathogenesis of isolated Pierre Robin sequence. *European Journal of Pediatrics, 161*, 275-280.

Aitkin, L. M., Byers M. & Nelson J. E. (1986). Brain stem auditory nuclei and their connections in a carnivorous marsupial, the northern native cat (Dasyurus hallucatus). *Brain Behavior and Evolusion, 29*, 1-16.

Anders, T. F., Keener, M. A. & Kraemer, H. (1985). Sleep-wake state organization, neonatal assessment and development in premature infants during the first year of life. II. *Sleep, 8*, 193-206.

Aston-Jones, G., Chen, S., Zhu, Y. & Oshinsky, M. L.(2001). A neural circuit for circadian regulation of arousal. *Nature Neuroscience, 4*, 732-738.

Aston-Jones, G. (2005). Brain structures and receptors involved in alertness. *Sleep Medicine, 6*, S3-7.

Atkinson, E., Vetere, A. & Grayson, K. (1995). Sleep disruption in young children. The influence of temperament on the sleep patterns of pre-school children. *Child: Care, Health and Development, 21*, 233-246.

Black, B., Holditch-Davis, D., Schwartz, T. & Scher, M. S. (2006). Effects of antenatal magnesium sulfate and corticosteroid therapy on sleep states of preterm infants. *Research in Nursing & Health, 29*, 269-280.

Borghese, I. F., Minard, K. L. & Thoman, E. B. (1995). Sleep rhythmicity in premature infants: implications for development status. *Sleep, 18,*523-530.

Brandon, D. H., Holditch-Davis, D. & Beylea, M. (1999). Nursing care and the development of sleeping and waking behaviors in preterm infants. *Research in Nursing and Health, 22,* 217-229.

Bremmer, P., Byers, J. F. & Kiehl, E. (2003) Noise and the premature infant: physiological effects and practice implications. *Journal of Obstetric, Gynecologic and Neonatal Nursing, 32*, 447-454.

Byers, J. F., Waugh, W. R. & Lowman, L. B. (2006). Sound level exposure of high-risk infants in different environmental conditions. *Neonatal Network, 25,* 25-32.

Curzi-Dascalova, L., Peirano, P. & Morel-Kahn, F. (1988). Development of sleep states in normal premature and fullterm newborns. *Developmental Psychobiology, 21*, 431–444.

Darnall, R. A., Ariagno, R. L., Kinney & H. C. (2006). The late preterm infant and the control of breathing, sleep, and brainstem development: a review. *Clinics in Perinatology, 33(4)*, 883-914.

Davidson, R. M. & Bender, D. B. (1991). Selectivity for relative motion in the monkey superior colliculus. *Journal of Neurophysiology, 65*, 1115-1133.

de Vries, J. I., Visser, G. H. & Prechtl, H. F. (1982). The emergence of fetal behaviour. I. Qualitative aspects. *Early Human Development, 7*, 301-322

Diambra, L. & Menna-Barreto, L. (2004). Infradian rhythmicity in sleep/wake ratio in developing infants. *Chronobiology International, 2*, 217-227.

Doljansky, J. T., Kannety, H. & Dagan, Y. (2005). Working under daylight intensity lamp: an occupational risk for developing circadian rhythm sleep disorder? *Chronobiology International, 22*, 597-605.

Feldman, R. (2004). Mother-infant skin-to-skin contact and the development of emotion regulation. S. P. Shohov (Ed.). *Advances in psychology research.* Vol. 27 (pp. 113-131). Hauppauge, NY; Nova Science.

Feldman, R. (2006). From biological rhythms to social rhythms: Physiological precursors of mother-infant synchrony. *Developmental Psychology, 42*, 175-88.

Feldman, R. (2007). Mother-infant contact and child development; Insights from the Kangaroo intervention. In L. L'Abate, D. D. Embry, & S. M. Baggett (Eds.) *Handbook of low-cost interventions to promote physical and mental health.* (pp. 323-351). New York; Springer

Feldman, R. & Eidelman, A. I. (2003). Mother-infant skin-to-skin contact (Kangaroo Care) accelerates autonomic and neurobehavioral maturation in premature infants. *Developmental Medicine and Child Neurology, 45*, 274-280.

Feldman, R., Weller, A., Sirota, L. & Eidelman, A. I. (2002). Skin-to-Skin contact (Kangaroo care) promotes self-regulation in premature infants: sleep-wake cyclicity, arousal modulation, and sustained exploration. *Developmental Psychology, 38,* 194-207.

Field, T. (1990). Alleviating stress in newborn infants in the intensive care unit. *Clinics in Perinatology, 17*, 1-9.

Field, T. (1995). Massage therapy for infants and children. *Journal of Behavior and Developmental Pediatrics, 16*, 105-111.

Frank, B., Frasch, M. G., Schneider, U., Roedel, M., Schwab, M. & Hoyer, D. (2006). Complexity of heart rate fluctuations in near-term sheep and human fetuses during sleep. *Biomedizinische Technik Biomedical engineering_,51*, 233-236.

Freudigman, K. A., Thoman, E. B. (1998). Infants' earliest sleep/wake organization differs as a function of delivery mode. *Developmental Psychobiology, 32*, 293-303.

Gonzalez, M. M. & Aston-Jones, G. (2006). Circadian regulation of arousal: role of the noradrenergic locus coeruleus system and light exposure. *Sleep. 1, 29*, 1327-1336.

Gossel-Symank, R., Grimmer, I., Korte, J. & Siegmund, R.(2004). Actigraphic monitoring of the activity-rest behavior of preterm and full-term infants at 20 months of age. *Chronobiology Internation, 21*, 661-671.

Halpern, L. F., Maclean, W. E. & Baumeister, A. A. (1995). Infant Sleep-Wake Characteristics: Relation to Neurological Status and the Prediction of Developmental Outcome, *Developmental Review, 15*, 255-291.

Harris, C. D. (2005). Neurophysiology of sleep and wakefulness. *Respiratory Care Clinics of North America, 11*, 567-586.

Hao H. & Rivkees, S. A. (1999). The biological clock of very premature primate infants is responsive to light. *Proceedings of the National Academy of Sciences of the United States of America, 2, 96*, 2426-2429.

High, P.C. & Gorski, P.A. (1985). Recording environmental influences on infant development in the intensive care nursery: Womb for improvement. In A.W. Gottfried & J.L. Gaiter (Eds.), Infant stress under intensive care: Environmental neonatology (pp. 131–155). Baltimore: University Park.

Hofer, M. A. (1995). Hidden regulators: Implications for a new understanding of attachment, separation, and loss. In S. Golberg, R. Muir, & J. Kerr (Eds.), *Attachment theory: Social, developmental, and clinical perspectives* (pp. 203-230). Hillsdale, NJ: Analytic Press.

Holditch-Davis, D. (1990). The development of sleeping and waking states in high-risk preterm infants. *Infant Behavior and Development, 13*, 513–531.

Holditch-Davis, D., Scher, M, Schwartz T. & Hudson-Barr D. (2004). Sleeping and waking state development in preterm infants. *Early Human Development, 80,* 43-64.

Ingersoll, E. W. & Thoman, E. B. (1999). Sleep/wake states of preterm infants: Stability, developmental change diurnal variation and relation with care-giving activity. *Child Development, 70*, 1-10.

Jiang, Z. D. & Wilkinson, A. R. (2006). Does peripheral auditory threshold correlate with brainstem auditory function at term in preterm infants? *Acta Oto-Laryngologica, 126*, 824-827.

Karmel, B. Z., Gardner, J. M. & Freedland, R. L. (1996). Arousal-modulated attention at four months as a function of intrauterine cocaine exposure and central nervous system injury. *Journal of Pediatric Psychology, 21*, 821-832.

Karmel, B. Z., Gardner, J. M., Zapulla, R. A., Magnano, C. L. & Brown, E. G. (1988). Brain-stem auditory evoked responses as indicators of early brain insult. *Electroencephalography and Clinical Neurophysiology, 71*, 429-442.

Keener, M. A., Zeanah, C. H., Anders, T. F. (1988). Infant temperament, sleep organization, and nighttime parental interventions. *Pediatrics, 81*, 762-771.

Kent, W. D., Tan, A. K., Clarke, M. C. & Bardell, T. (2002). Excessive noise levels in the neonatal ICU: potential effects on auditory system development. *Journal of Otolaryngology, 31*, 355-60.

Kleitman, N. (1957). Sleep, wakefulness and consciousness. *Psychological Bulletin, 54*, 354-359.

Kopp, C., Vogel, E., Rettori, M. C., Delagrange, P., Renard, P., Lesieur, D. & Misslin, R. (1999) Regulation of emotional behaviour by day length in mice: implication of melatonin. *Behavioral Pharmacology,10*, 747-752.

Korte, J., Hoehn, T. & Siegmund, R. (2004). Actigraphic recordings of activity-rest rhythms of neonates born by different delivery modes. *Chronobiology International, 21*, 95-106.

Krueger, C., Wall, S., Parker, L. & Nealis, R. (2005). Elevated sound levels within a busy NICU. Neonatal Network, 24, 33-37.

Lai, Y. Y., Shalita, T., Hajnik, T., Wu, J.P., Kuo, J. S., Chia, L. G. & Siegel J. M. (1999). Neurotoxic N-methyl-D-aspartate lesion of the ventral midbrain and mesopontine junction alters sleep-wake organization. *Neuroscience, 90*, 469-483.

Larson, C. R., Yajima, Y. & Ko, P. (1994). Modification in activity of medullary respiratory-related neurons for vocalization and swallowing. *Journal of Neurophysiology,71*, 2294-2304.

Leitner, Y., Bloch, A. M., Sadeh, A., Neuderfer, O., Tikotzky, L., Fattal-Valevski, A. & Harel, S. (2002). Sleep-wake patterns in children with intrauterine growth retardation. *Journal of Child Neurology, 17*, 872-876.

Losier, B. J. & Semba K. (1993). Dual projections of single cholinergic and aminergic brainstem neurons to the thalamus and basal forebrain in the rat. *Brain Research, 26, 604*, 41-52.

McGraw, K., Hoffmann, R., Harker, C. & Herman, J. H. (1999). The development of circadian rhythms in a human infant. *Sleep, 1 , 22*, 303-310.

McMillen, I. C., Kok, J. S., Adamson, T. M., Deayton, J. M. & Nowak R. (1991). Development of circadian sleep-wake rhythms in preterm and full-term infants. *Pediatric Research, 29,* 381-384.

McNamara, F., Wulbrand, H. & Thach, B. T. (1999). Habituation of the infant arousal response. *Sleep, 1, 22*, 320-326.

Mirmiran, M. & Ariagno, R. L. (2000). Influence of light in the NICU on the development of circadian rhythms in preterm infants. *Seminars in Perinatology, 24*, 247-257.

Mirmiran, M., Baldwin, R. B. & Ariagno, R. L. (2003). Circadian and sleep development in preterm infants occurs independently from the influences of environmental lighting. *Pediatric Research, 53*, 933-938.

Nattie, E. (2000). Multiple sites for central chemoreception: their roles in response sensitivity and in sleep and wakefulness. *Respiration Physiology, 122*, 223-235.

Niessen, F. (2006). [Neurosensory visual development of the foetus and newborn and neonatal intensive care units]. *Archives de Pediatrie, 13*, 1178-1184. [Article in French]

Nikkol, E. M., Kirjavainen, T. T., Ekblad, U. U., Kero, P. O. & Salonen, M. A. (2002). Postnatal adaptation after Caesarean section or vaginal delivery, studied with the static-charge-sensitive bed. *Acta Paediatrica, 91*, 927-933.

Olischar, M., Klebermass, K., Kuhle, S., Hulek, M., Kohlhauser, C., Rucklinger, E., Pollak, A. & Weninger, M. (2004). Reference values for amplitude-integrated electroencephalographic activity in preterm infants younger than 30 weeks' gestational age. *Pediatrics, 113*, e61-66.

Parmelee, A. H., Jr., Wenner, W. H., Akiyama, Y., Schultz, M. & Stern, E. (1967). Sleep states in premature infants. *Developmental Medicine and Child Neurology, 9*, 70–77.

Philbin, M. K., Ballweg, D. D. & Gray, L. (1994). The effect of an intensive care unit sound environment on the development of habituation in healthy avian neonates. *Developmental Psychobiology, 27*, 11-21.

Rivkees, S. A. (2003). The management of hyperthyroidism in children with emphasis on the use of radioactive iodine. *Pediatric Endocrinology Reviews, 1*, 212-21.

Rivkees, S. A., Mayes, L., Jacobs, H. & Gross, I. (2004). Rest-activity patterns of premature infants are regulated by cycled lighting. *Pediatrics, 113*, 833-839.

Sadeh, A., Dark, I. & Vohr, B. R. (1996). Newborns' sleep-wake patterns: the role of maternal, delivery and infant factors. *Early Human Development, 23*, 44, 113-126.

Scher, M. S., Johnson, M. W. & Holditch-Davis, D. (2005). Cyclicity of neonatal sleep behaviors at 25 to 30 weeks' postconceptional age. *Pediatric Research, 57*, 879-882.

Shimada, M., Takahashi, K., Segawa, M., Higurashi, M., Samejim, M. & Horiuchi, K. (1999). Emerging and entraining patterns of the sleep-wake rhythm in preterm and term infants. *Brain & Development, 21*, 468-473.

Shimada, M., Segawa, M., Higurashi, M., Kimura, R., Oku, K., Yamanam,i S. & Akamatsu, H. (2003). Effects of phototherapy in neonates on circadian sleep-wake and saliva cortisol level rhythms. *The Journal of Perinatal and Neonatal Nursing, 17*, 222-231.

Sleigh, M. J. & Lickliter, R. (1998). Timing of presentation of prenatal auditory stimulation alters auditory and visual responsiveness in bobwhite quail chicks (Colinus virginianus). *Journal of Comparative Psychology, 112*, 153-60.

Torres-Farfan, C., Rocco, V., Monso, C., Valenzuela, F. J., Campino, C., Germain, A., Torrealba, F., Valenzuela, G. J. & Seron-Ferre, M. (2006). Maternal melatonin effects on clock gene expression in a nonhuman primate fetus. *Endocrinology, 147*, 4618-4626.

Van Leeuwen, P., Cysarz, D., Lange, S. & Gronemeyer, D. (2006). Increase in regularity of fetal heart rate variability with age. *Biomedizinische Technik Biomedical engineering, 51*, 244-247.

Wong, V. & Wong, S. N. (1991). Brainstem auditory evoked potential study in children with autistic disorder. *Journal of Autism and Developmental Disorders, 21*, 329-340.

Yilmaz, Y., Degirmenci, S., Akdas, F., Kulekci, S., Ciprut, A., Yuksel, S., Yildiz, F., Karadeniz, L. & Say, A. (2001). Prognostic value of auditory brainstem response for neurologic outcome in patients with neonatal indirect hyperbilirubinemia. *Journal of Child Neurology, 16*, 772-775.

In: Biological Clocks: Effects on Behavior... ISBN: 978-1-60741-251-9
Editors: O. Salvenmoser et al. pp. 121-137

Chapter 5

INTERACTIONS BETWEEN THE CIRCADIAN AND THE IMMUNE SYSTEM: A FRAMEWORK FOR THE UNDERSTANDING OF DISEASE

Natalia Paladino, María Juliana Leone, Leandro P. Casiraghi, Patricia V. Agostino, Diego A. Golombek* and Juan J. Chiesa

Chronobiology Laboratory, National University of Quilmes/CONICET, Buenos Aires, Argentina

ABSTRACT

The daily environmental changes have imposed a selective pressure for life on Earth, driving the development of a circadian clock mechanism for the generation and entrainment of rhythms in physiological and behavioral variables (e.g. body temperature, hormonal secretion, sleep, locomotor activity, etc.). In mammals, the clock resides in the hypothalamic suprachiasmatic nuclei (SCN), and the principal signal that adjusts its activity is the light-dark cycle. Most immune factors and processes are under circadian control, although the efferent pathways that control these cycles are not completely understood. In addition, circadian disorders are usually associated with disease, and temporal disarrangements in immune parameters

---

*Correspondence Author: Dr. Diego Golombek, Departamento de Ciencia y Tecnología, Universidad Nacional de Quilmes, R.S. Peña 352, (1876) Bernal, Buenos Aires, Argentina. Fax +54-11-4365-7132. E-mail: dgolombek@unq.edu.ar

are closely related to the onset or the development of pathologies. In particular, several lines of evidence link the circadian clock to cancer disease. Epidemiologic studies show associations between circadian rhythms disruption and cancer risk or inadequate responses to treatment. Moreover, proinflammatory cytokines, some of which have been reported to be key actors in tumour signaling, can generate deep behavioral alterations comprising clock-associated variables during tumor progression, such as altered sleep pattern, fatigue, and lower quality of life.

We argue that the circadian-immune interaction operates in two directions: while the SCN (or other, peripheral, clocks) might drive circadian variations in several immune variables, humoral signals can in turn affect the molecular mechanism of the circadian clock and its entrainment. For example, higher diurnal lethality depicts a strong circadian modulation of lipopolysaccharide (LPS)-induced sepsis, which is also observed using TNF-α stimulation. Current clinical data show a strong correlation between time of day and illness manifestation or immune activity. In the other direction, administration of subpyrogenic doses of LPS, as well as of proinflammatory cytokines IL-1β and TNF-α induces changes in SCN-driven locomotor activity rhythms. At the molecular level, this effect is achieved through activation of a transcriptional pathway involving nuclear factor-κB (NF-κB). This and other pathways (such as mitogen-activated-kinase (MAPK) signaling), lead to the expression of specific genes of the circadian clockwork, which in turns regulates its rhythmic output. Regarding cancer disease, the clock genes *Per1* and *Per2* are involved in DNA damage response, and could therefore regulate tumour suppression. Moreover, the expression of many cell-cycle genes is under circadian control. Alterations in other components of the molecular clock, as those in members of the casein kinase (CK) family, have also been linked to cancer disease.

In this review we analyze the interaction between the circadian and immune systems, and explore the signal transduction pathways which could participate in this dialogue. The precise knowledge of this interaction might be extremely useful for the understanding of diseases development such as cancer.

INTRODUCTION

Circadian Rhythms in the Immune System

Circadian systems in mammals are hierarchical structures of clocks and oscillators in cells and tissues, whose primary functions are to provide temporal organization of physiology and behavior, and to enable the organism to anticipate regular changes in their environment [1]. At the top of this hierarchy are the suprachiasmatic nuclei (SCN), which coordinate and synchronize circadian rhythms. Within the SCN, rhythms are generated in circadian pacemaker cells by a complex of molecular feedback loops that positively and negatively regulate the transcription of core genes (e.g. *per, cry, bmal1*) of the circadian clock [2]. Rhythms are ubiquitous for most physiological and behavioral variables, including body temperature, sleep, hormonal secretion and locomotor activity. The activity of immune system is not an exception, exhibiting a strong daily modulation. Several lines of evidence suggest a crosstalk between the circadian and the immune system (reviewed in [3]). In early experiments, a circadian variation in susceptibility to lethal doses of bacterial lipopolysaccharide (LPS) endotoxin was demonstrated in rodents, with higher mortality occurring during the day [4]. Indeed, this day/night variation in lethality was observed in a more recent experiment using tumor necrosis factor-α (TNF-α), one of the main LPS-induced proinflammatory cytokine [5]. These findings show that circadian regulation of immune-related variables determine daily variations in the intensity of inflammatory processes, not only in response to LPS stimuli, but also in the course of autoimmune diseases [6]. For instance, in humans, the number of lymphocytes, eosinophils and basophile granulocytes is higher during the night, whereas monocytes and neutrophiles fall during the day. In addition, major humoral immune responses undergo circadian changes, as for instance antibodies titer after immunization, or peak concentration of three major classes of immunoglobulins (reviewed in [7]). Either rhythms in plasmatic levels of pro-inflammatory cytokines, as well as peptide hormones produced and secreted by immune cells, were also reported. Interleukin-1β (IL-1β), 6, 10 and 12, TNF-α, and interferon-γ (IFN-γ) exhibit in most cases high levels during the night and early morning in the human [8]. The rise in cytokine plasmatic levels may at least in part explain the exacerbation of immuno-inflammatory disorders at these times of the day. Furthermore, the fact that nocturnal rodents exhibit rhythms with opposite phase to diurnal ones suggests a relationship between behavioral state and immune activity.

Circadian-Immune Cross-Talk

The circadian control of the immune system might involve SCN efferents towards both neuroendocrine and autonomic neurons of the paraventricular nuclei (PVN) and the sub-PVN region in the hypothalamus [9]. Neurons projecting from these centers towards sympathetic and parasympathetic systems send circadian information to different organs, including the pineal and adrenal glands, the liver, etc. Not only primary and secondary lymphoid organs are richly supplied with autonomic (mainly sympathetic) innervations [10], but also the peak number of immune cells may be temporally regulated by the autonomic nervous system, by means of an increased adrenergic/cholinergic receptor proportion in those peaking during the day; the opposite occurs in those that peak at night [11]. The main diurnal immunomodulators are likely to be corticosteroids, which are under strong circadian control. SCN projections on hypothalamic neurons that release corticotrophin-releasing hormone stimulate pituitary adrenocorticotrophin hormone, leading to corticosteroid secretion. Thus, the circadian activation of the hypothalamic-pituitary-adrenal (HPA) axis produces a morning peak in plasmatic corticosteroids level to enhance stress responses in the beginning of activity. In addition, the immune system feedbacks on autonomic abdominal vagus nerves through the paracrine action of cytokines, which interact with receptors in vagal sensory neurons [12]. Changes in immune function can also influence the expression of neural receptors in lymphoid organs [10]. The activation of the HPA axis, leading to increased corticosteroids, occurs in response to pro-inflammatory cytokines (especially IL-1) secreted at the site of injury, or in the course of immune challenges, promoting anti-inflammatory responses in a feedback regulatory mechanism [7]. Thus, the immune system acts through the production of cytokines, which signal the exposure to antigen to the central nervous and endocrine systems. In addition, extensive data support the idea that immunological variables are influenced by the amount or distribution of sleep, gated by circadian clock. For instance, natural killer (NK) cell activity, antigen uptake, circulating immune complexes, secondary antibody responses are altered by sleep loss [13], and also IL-1β and TNF-α levels increase during sleep deprivation [14].

Not only there is a clear circadian influence on mammalian immune function; conversely, there is also data showing that circadian timing mechanisms are under the influence of the immune system, raising the likelihood that there is a bidirectional communication between the circadian and immune systems. Some *in vivo* experiments indicate this interaction, in which high doses of LPS or pathological states (sepsis or infection with blood-borne parasites) severely affect the circadian output [4, 15-18, unpublished results from our laboratory]. Also, we found that systemic low doses of endotoxin lipopolysaccharide (LPS) delivered at Circadian Time (CT) 15 (about 3h after locomotor activity onset) induces phase-delays of locomotor activity rhythms in mice [19]. This result was the first *in vivo* evidence of the response of the circadian system to an immune challenge, and suggests the existence of a feedback pathway from immune effectors into the circadian system. Moreover, i.c.v. injections of IL-1β and TNF-α, which are strongly stimulated by LPS, also induce phase-delays at CT15 (unpublished results), suggesting that LPS may act on the clock by inducing the release of cytokines which finally act on the SCN. Futhermore, the sleep–wake cycle (one of the most evident circadian rhythms) is modified by cytokines (with proinflammatory cytokines exerting a somnogenic effect [13, 14]). Also, since immune suppression with Ciclosporin A (CsA) or FK-506 has no effect on clock phase at CT15, but do induces phase changes at other times of the day, an opposite effect of immune activating or inhibiting factors might be suggested [20].

Immune factors can affect the circadian clock at the molecular oscillator mechanism. It was shown that LPS modifies neurochemical SCN activity in organotypic slices, increasing arginine-vasopressin secretion [21]. Also, acute-phase proinflammatory factors can stimulate clock-genes expression, such as IL-6 enhancing *Per1* in human hepatoma (HuH-7) cells [22]. The rhythmicity of clock genes and the photic induction of *Per1* in SCN were also disturbed by the repetitive administration of IFN-α at CT15 [23]. In addition, there is evidence for a role of NF-κB, a classical immune transcription factor, in this interaction. Indeed, we verified the expression and activity of this factor at the SCN and found that NF-κB inhibition blocks both LPS- and light-induced phase shifts of circadian rhythms [19, 24]. Since SCN has a dense and rhythmic immunorreactivity for Glial Fibrillary Acidic Protein (GFAP, specific marker of astroglial cells), we hypothesized that astroglia could be the cellular substrate for this interaction. We have found NF-κB expression in SCN astrocytes *in vitro* and in brain slices; in addition, NF-κB was activated in response to LPS and proinflammatory cytokines in SCN astroglial cultures [25].

In addition to NF-κB, the MAPK and NO-cGMP are two common pathways involved both in the circadian and immune systems function. The circadian fluctuation of MAPK activity in the SCN is involved in both input and output SCN pathways and has an essential role in the maintenance of robust circadian clock gene expression [26] and in the light-induced response of SCN neurons [27]. The MAPK signaling pathway also plays a central role in the inflammatory cytokine response to immune challenge. Many effects of proinflamatories cytokines IL-1, IL-6 or TNF-α, including up-regulation of chemokines, cytokines and adhesion molecules involve MAPK transduction pathways [28-31].

On the other hand, the NO-cGMP pathway is involved in light-induced phase advances of the circadian rhythms [32, 34]. cGMP levels in the hamster SCN exhibit circadian variations with maximum values during the subjective day. This variation appears to be related to temporal changes in cGMP phosphodiesterase (PDE) activity and not to guanylyl cyclase (GC) activity. Light pulses increase SCN cGMP levels at CT18 (when light causes phase advances of rhythms) but not at CT 14 (the time for light-induced phase delays). cGMP-dependent kinase II (cGK II) is expressed in the hamster SCN and also exhibits circadian changes in its levels, peaking during the day. Light pulses increase cGK II activity at CT18, but not at CT14, and affect the expression of the clock gene *Per* in the SCN [32]. Similar to MAPK, the NO-cGMP pathway is also modulated by proinflammatory cytokines. It has been reported that IL-1β or TNF-α down-regulated the NO-dependent cGMP synthesis by inhibiting GC activity [35, 36]. Although the precise effects of cytokines in MAPK or NO-cGMP pathways in the SCN have not been explored, it could represent a possible mechanism for the immune system modulation of circadian rhythms.

Chronobiological Factors in Tumor Development

A major consequence of contemporary life is the disruption of circadian rhythms, leading to a number of pathological conditions, including sleep disturbances and depression. The circadian clock-cancer connection has been investigated in studies of airline pilots, flight attendants and shift workers, who are more likely to have disrupted circadian cycles due to abnormal working hours [37]. Disruption of circadian rhythms in mice generated by SCN lesions [38, 39] or by chronic exposure to experimental jet lag [40] is associated with accelerated growth of malignant tumors. In fact, it has been documented that both the circadian cortisol and activity rhythms robustness are prognostic factors for survival in patients with colorectal cancer [41, 42].

Incidence of hormone-related breast and prostate cancer increases significantly in women and men, respectively, working nightshifts [43, 44]. Moreover, prognosis of cancer is poorer in patients with altered circadian rhythms compared to patients with normal rhythms [43]. This association between disrupted circadian cycles and the increased rates of cancer may be due to circadian hormonal disturbances [45]. Indeed, disruption of circadian melatonin rhythms in rodents, either by pinealectomy or through prolonged exposure to constant light, increases the formation of spontaneous mammary tumors. When acting as a chronobiotic (i.e., modifying circadian system), melatonin has demonstrated significant antitumoral effects on tumor cell lines and in a melanoma-murine model [reviewed in 37]. However, despite the accumulating evidence for a causal link between disruption of circadian cycle and tumorigenesis, the molecular mechanisms underlying this effect remain poorly understood.

As we described previously, circadian rhythms have been observed in immune system components, including cytokine levels and NK or T cells number, which are essential mediators of the antitumoral response [46]. Disruption of the circadian system may result in disordered immune responses such as aberrant immune cell trafficking and abnormal cell proliferation cycles [47]. Sleep disorders, which result in an altered circadian rhythm, have been shown to affect the activation of the HPA axis, a major neuroendocrine component regulating the immune system [48]. This change in HPA activity consequently affects the expression of proinflammatory cytokines, such as IL-6 or TNF-α. In line with this concept, IL-6 and other cytokines has been shown to correlate with the expression of several cancer types [49]. A comparative study in colon cancer patients with pronounced or dampened circadian rhythms, showed that increased IL-6 and TGF-α levels in the dampened group are associated with poorer performance status, impaired emotional and social functioning, worsened appetite loss, increased fatigue and disrupted circadian motor activity. This finding is of particular interest in light of the recent reports in rodents that TGF-α, highly expressed in SCN, mediates hypothalamic signaling for the circadian regulation of motor activity, sleep, and body temperature. This signaling occurs through activation of multiple efferents including the epidermal growth factor (EGF) receptor-positive staining cells in the subparaventricular zone [50]. Moreover, infusion of TGF-α into this neural zone blocks signaling from the SCN [51].

Circadian Alterations Induced by Cancer

Cancer patients frequently have fatigue, loss of appetite, change in sleep patterns, and other symptoms that are assessed by quality of life indexes and performance status scores. The importance of these data is illustrated by low scores predicting for treatment failure and poor survival in several types of cancer. The quality of life indexes have been shown to be correlated with clinical measurements of circadian rest/activity patterns [42]. Behavioral changes in cancer patients are similar to those observed in the ''sickness behavior'' animal model, which display physiologic components of an acute phase response with fever, pain, social withdrawal, decreased food intake, and wasting. These systemic symptoms are mediated by circulating cytokines and/or by those centrally produced by afferent nerve fibers, through a cascade of the proinflammatory cytokines IL-1, TNF-α, and IL-6, that signals to the HPA axis and the autonomic nervous system [52, 53]. Further preclinical evidence strengthened the putative role of cytokines in sickness behavior, supported by the finding of cytokine receptors for TGF-α, EGF, TNF-α, and IL-6 in many regions of the brain [42]. Moreover, administration of IFN-α and IL-2 in humans is associated with the development of cognitive and depressive symptoms [54]. Interestingly, an inhibitory effect on mRNA expression of each clock gene in the SCN was observed in humans and rodents by the repetitive administration of IFN-α during the early active phase, but not during the early rest phase [55]. Taking together all this evidences, it is plausible that cytokines mediate the dialogue between the immune and the circadian systems during sickness behavior. Moreover, this immune-circadian interaction could be able to modulate the individual immune response to pathologic state, as cancer disease, or the success of different treatment regimens.

Implications of Molecular Clock Components in Cancer

It has been estimated that 2–10% of all mammalian genes are clock-controlled genes, indicating an extensive circadian gene regulation. In particular, the molecular clock components can modify the transcriptional activity of some key genes involved into cell cycle regulation, thereby suggesting that the circadian clock could regulate cell proliferation [56]. Recent studies have identified *c-myc, p53* and *wee1* as being clock-controlled genes. *C-myc* and *wee1* promote cell cycle progression from G1 to S, and from G2 to M, respectively. Furthermore, *c-myc* can also exert pro-apoptotic effects through p53-dependent or independent pathways. Many other genes that control cell cycle progression or apoptosis display 24-h rhythms in mRNA and/or protein expression in healthy tissues obtained from rodents and/or humans [reviewed in 57].

Molecular clock components have also been involved in tumorigenesis. The circadian genes *Per1* and *Per2* appear to play an important role in tumor suppression *in vivo* [58, 59]. Also, the protein kinases CK1ε y δ (which are able to regulate the circadian period by phosphorylating clock proteins PER and CRY, driving proteosome-mediated degradation [60, 61]) can also phosphorylate key proteins in cancer signaling pathways, such as p53 [62] and β-catenin [63]. Microarray expression data for CK1ε from different tumoral tissues (including brain, renal, leukemia, lung, melanoma, prostate, and seminoma), show upregulated expression compared to normal tissues. Additionally, knocking down CK1ε expression induces apoptotic cell death, and it also generates cell cycle arrest in sensitive cancer cells [64]. Furthermore, there is evidence suggesting that CK1ε exerts its oncogenic effects by inhibiting the tumor suppressive function of

PER2 [64]. Similarly, CK2, an important regulator of circadian rhythm in *Drosophila melanogaster* by phosphorilation of PER protein [65] has also been associated to cancer disease in mammals [66].

Importantly, other circadian regulators appear to play roles in cell cycle control and DNA damage response. PER1 interacts with cell cycle checkpoint proteins ATM and CHK2, [59] whereas the human Timeless protein interacts with both CRY2 and checkpoint proteins CHK1 and the ATR-ATRIP complex [67]. Despite the recent developments linking clock-controlled proteins and cell cycle, it is not known whether these checkpoint proteins are modified by interactions with core clock proteins leading to alteration in their activities.

It has been shown that CLOCK-BMAL1 clock proteins directly regulate cell cycle genes such as *wee1* (G2/M transition) [68], *c-myc* (G0/G1 transition) and *Cyclin* D1 (G1/S transition) [58]. Additionally, CLOCK has an intrinsic histone acetyltransferase (HAT) activity [69], by virtue of which it can induce chromatin remodeling and establish a permissive state for activation of gene expression. Other HATs, such as CBP and p300, are known to acetylate also non-histone targets and have been recognized as tumor suppressors. Similarly, CLOCK could also target non-histone proteins. Although the HAT function has not been linked to regulation of cell proliferation yet, it is likely that CLOCK plays an important role because of its control of circadian physiology and cellular metabolism [45].

Conclusion

We have presented a set of evidences describing the circadian-immune interaction in two directions. On one hand, although most immune factors and processes are under daily control, orchestrated by the circadian pacemaker located at the hypothalamic SCN system, the efferent pathways that control these cycles are not completely understood. On the other hand, these same temporal variations might be useful for a fine-tuning of the system by affecting receptors located in the circadian clock. Importantly, circadian disorders are usually associated with disease, and temporal disarrangements in immune parameters are closely related to the onset or development of pathological mechanisms. We could speculate on the existence of a feedback loop in which immune factors, mainly proinflammatory cytokines, mediate the clock-cancer interaction. Briefly, an abnormal regulation of the circadian rhythm in immune parameters might collaborate with the generation of cancerous processes or affect the outcome of disease. Conversely, tumor development induces the expression and/or activity of proinflammatory cytokines, which could affect circadian clock function and promote sickness behavior (Figure 1). Thus, alteration of the circadian system may be overcome by devising a dosing regime that minimizes adverse drug effects on clock function.

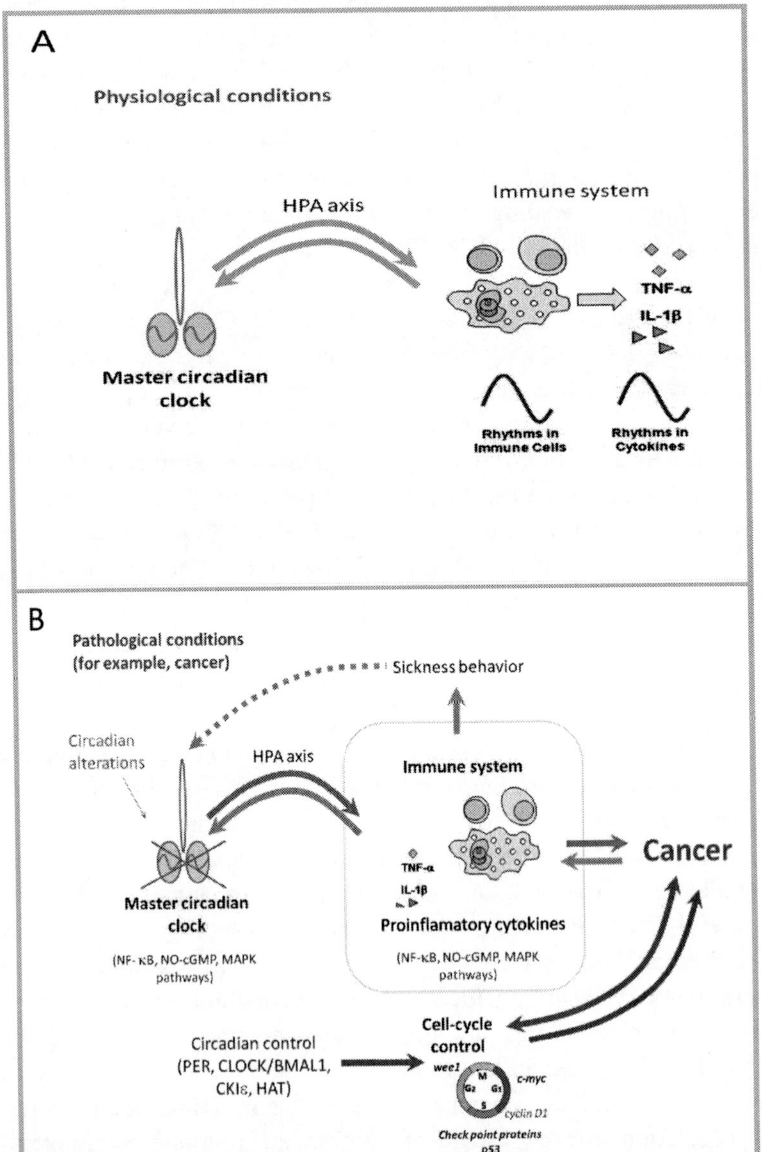

Figure 1. General interactions between the immune and circadian systems, in physiological and pathological conditions (like cancer). In physiological conditions (A), the master

circadian clock is entrained and controls global rhythms, such as cycles in circulating immune cells number and cytokine levels. Some immune factors feedback into the circadian clock providing signals about the physiological state of the organism. Under exogenous alterations (e.g., jet lag, shift work), the clock is affected producing altered circadian rhythms. Alterations in rhythms of the immune system can be related to a higher incidence of cancer (B, green arrows). When some cells become cancerous, the immune system is activated and several proinflammatory cytokines are released (IL-1, TNF-α, among others). These factors are responsible for sickness behavior and can also affect clock activity (blue arrows). Cancer is directly related with changes in the normal cellular cycle, through some key factors that induce an abnormally high division rate (red arrow). The molecular circadian clock can regulate the expression of several cell cycle-factors, affecting cell phenotype (violet arrows). In addition, cytokines can modify clock gene expression in several cell types (modified from [70]).

Besides a better understanding of the physiological regulation of the circadian-immune interaction, its chronobiological basis will therefore provide a novel framework for diagnosis and therapeutical approaches in order to treat disease. Moreover, the precise knowledge of the circadian regulation of immune factors in the context of tumorigenesis might be extremely useful for the understanding of tumor progression, or to prevent circadian alterations (including the sleep-wake cycle) which greatly decrease the patients' quality of life.

REFERENCES

Pittendrigh, C. S. (1981). In: *Handbook of Behavioral Neurobiology, Biological Rhythms, J. Aschoff, Ed.* (Plenum Press, New York), pp. 95-124.

Ko, C. H. & Takahashi, J. S. (2006). Molecular components of the mammalian circadian clock. *Hum. Mol. Genet., 15*, R271-277.

Coogan, A. N. & Wyse, C. A. (2008). Neuroimmunology of the circadian clock. *Brain Res, 1232*, 104-12.

Halberg, F., Johnson, E. A., Brown, B. W. & Bittner, J. J. (1960). Susceptibility rhythm to E. coli endotoxin and bioassay. *Proc Soc Exp Biol Med, 103*, 142-144.

Hrushesky, W. J., Langevin, T., Kim, Y. J. & Wood, P. A. (1994). Circadian dynamics of tumor necrosis factor α (cachectin) lethality. *J Exp Med, 180(3)*, 1059-65.

Cutolo, M. & Masi, A. T. (2005). Circadian rhythms and arthritis. *Rheum Dis Clin North Am, 31*, 115-29.

Haus, E. & Smolensky, M. H. (1999). Biologic rhythms in the immune system.

Chronobiol Int, 16, 581-622.

Petrovsky, N. & Harrison, L. C. (1997). Diurnal rhythmicity of human cytokine production: a dynamic disequilibrium in T helper cell type 1/T helper cell type 2 balance? *J Immunol, 158*, 5163-5168.

Buijs, R. M, van Eden, C. G, Goncharuk, V. D. & Kalsbeek, A. (2003). The biological clock tunes the organs of the body: timing by hormones and the autonomic nervous system. *J Endocrinol, 177*, 17-26.

Mignini, F, Streccioni, V. & Amenta, F. (2003). Autonomic innervation of immune organs and neuroimmune modulation. *Auton Autacoid Pharmacol, 23*, 1-25.

Suzuki, S., Toyabe, S., Moroda, T., Tada, T., Tsukahara, A., Iiai, T., Minagawa, M., Maruyama, S., Hatakeyama, K., Endoh, K. & Abo, T. (1997). Circadian rhythm of leucocytes and lymphocytes subsets and its possible correlation with the function of the autonomic nervous system. *Clin Exp Immunol, 110*, 500-508.

Konsman, J. P, Parnet, P. & Dantzer, R. (2002). Cytokine-induced sickness behaviour: mechanisms and implications. *Trends Neurosci, 25*, 154-159.

Krueger, J. M, Majde, J. A. & Obal, F. (2003). Sleep in host defense. *Brain Behav Immun, 17* Suppl 1, S41-47.

Majde, J. A. & Krueger, J. M. (2005). Links between the innate immune system and sleep. J Allergy *Clin Immunol, 116*, 1188-1198.

Bauhofer, A., Witte, K., Celik, I., Pummer, S., Lemmer, B. & Lorenz, W. (2001). Sickness behaviour, an animal equivalent to human quality of life, is improved in septic rats by G-CSF and antibiotic prophylaxis. *Langenbecks Arch Surg, 386*, 132-140.

Bauhofer, A., Witte, K., Lemmer, B., Middeke, M., Lorenz, W. & Celik, I. (2002). Quality of life in animals as a new outcome for surgical research: G-CSF as a quality of life improving factor. *Eur Surg Res, 34*, 22-29.

Bentivoglio, M., Grassi-Zucconi, G., Peng, Z. C. & Kristensson, K. (1994). Sleep and timekeeping changes, and dysregulation of the biological clock in experimental trypanosomiasis. *Bull Soc Pathol Exot* 87, 372-375.

Ebong, S., Call, D., Nemzek, J., Bolgos, G., Newcomb, D. & Remick, D. (1999). Immunopathologic alterations in murine models of sepsis of increasing severity. *Infect Immun, 67*, 6603-6610.

Marpegan, L., Bekinschtein, T. A., Costas, M. A. & Golombek, D. A. (2005). Circadian responses to endotoxin treatment in mice. *J Neuroimmunol, 160*, 102-109.

Katz, M. E, Simonetta, S. H, Ralph, M. R. & Golombek, D. A. (2008). Immunosuppressant calcineurin inhibitors phase shift circadian rhythms and inhibit circadian responses to light. *Pharmacol Biochem Behav.*, (doi: 10.1016/j.pbb.2008.05.017).

Nava, F., Carta, G. & Haynes, L. W. (2000). Lipopolysaccharide increases argininevasopressin release from rat suprachiasmatic nucleus slice cultures. *Neurosci Lett, 288*, 228-230.

Motzkus, D., Albrecht, U. & Maronde, E. (2002). The human PER1 gene is inducible by interleukin-6. *J Mol Neurosci, 18*, 105-109.

Ohdo, S., Koyanagi, S., Suyama, H., Higuchi, S. & Aramaki, H. (2001). Changing the dosing schedule minimizes the disruptive effects of interferon on clock function. *Nat Med., 7(3)*, 356-60.

Marpegan, L., Bekinschtein, T. A., Freudenthal, R., Rubio, M. F., Ferreyra, G. A., Romano, A. & Golombek, D. A. (2004). Participation of transcription factors from the Rel/NFkappa B family in the circadian system in hamsters. *Neurosci Lett, 358*, 9-12.

Leone, M. J, Marpegan, L., Bekinschtein, T. A., Costas, M. A. & Golombek, D. A. (2006). Suprachiasmatic astrocytes as an interface for immune-circadian signalling. *J Neurosci Res, 84*, 1521-1527.

Akashi, M., Hayasaka, N., Yamazaki, S. & Node, K. (2008). Mitogen-activated protein kinase is a functional component of the autonomous circadian system in the suprachiasmatic nucleus. *J Neurosci., 28(18)*, 4619-23.

Hainich, E. C, Pizzio, G. A. & Golombek, D. A (2006). Constitutive activation of the ERK-MAPK pathway in the suprachiasmatic nuclei inhibits circadian resetting. *FEBS Lett., 580(28-29)*, 6665-8.

Elsea, C. R, Roberts, D. A, Druker, B. J. & Wood, L. J. (2008). Inhibition of p38 MAPK suppresses inflammatory cytokine induction by etoposide, 5-fluorouracil, and doxorubicin without affecting tumoricidal activity. *PLoS ONE, 3(6)*, e2355.

Ho, A. W., Wong, C. K. & Lam, C. W. (2008). Tumor necrosis factor-alpha up-regulates the expression of CCL2 and adhesion molecules of human proximal tubular epithelial cells through MAPK signaling pathways. *Immunobiology, 213(7)*, 533-44.

Chowdhury, T. T., Salter, D. M., Bader, D. L. & Lee, D. A. (2008). Signal transduction pathways involving p38 MAPK, JNK, NFkappaB and AP-1 influences the response of chondrocytes cultured in agarose constructs to IL-1beta and dynamic compression. *Inflamm Res., 57(7)*, 306-13.

Kang, M. K. & Kang, S. K. (2008). Interleukin-6 induces proliferation in adult spinal cord-derived neural progenitors via the JAK2/STAT3 pathway with EGF-induced MAPK phosphorylation. *Cell Prolif., 41(3)*, 377-92.

Ferreyra, G. A. & Golombek, D. A. (2001). Rhythmicity of the cGMP-related signal transduction pathway in the mammalian circadian system. *Am J Physiol Regul Integr Comp Physiol., 280(5)*, R1348-55.

Golombek, D. A., Agostino, P. V., Plano, S. A. & Ferreyra, G. A. (2004). Signaling in the mammalian circadian clock: the NO/cGMP pathway. *Neurochem Int., 45(6)*,929-36.

Agostino, P. V, Plano, S. A. & Golombek, D. A. (2007). Sildenafil accelerates reentrainment of circadian rhythms after advancing light schedules. *Proc Natl Acad Sci U S A., 104*, 9834-9.

Pedraza, C. E, Baltrons, M. A, Heneka, M. T. & García, A. (2003). Interleukin-1 beta and lipopolysaccharide decrease soluble guanylyl cyclase in brain cells: NO-independent destabilization of protein and NO-dependent decrease of mRNA. *J Neuroimmunol., 144*, 80-90.

Browner, N. C, Sellak, H. & Lincoln, T. M. (2004). Downregulation of cGMP-dependent protein kinase expression by inflammatory cytokines in vascular smooth muscle cells. *Am J Physiol Cell Physiol., 287(1)*, C88-96.

Zhu, Y., Zheng, T., Stevens, R. G., Zhang, Y. & Boyle, P. (2006). Does ''Clock'' Matter in Prostate Cancer? *Cancer Epidemiol Biomarkers Prev., 15(1)*, 3-5.

Filipski, E., King, V. M., Li, X., Granda, T. G, Mormont, M. C., Liu, X., Claustrat, B., Hastings, M. H. & Lévi, F. (2002). Host circadian clock as a control point in tumor progression. *J Natl Cancer Inst, 94*, 690-7.

Otálora, B. B, Madrid, J. A, Alvarez, N., Vicente, V. & Rol, M. A. (2008). Effects of exogenous melatonin and circadian synchronization on tumor progression in melanoma-bearing C57BL6 mice. *J Pineal Res., 44(3)*,307-15.

Filipski, E., Delaunay, F., King, V. M., Wu, M. W., Claustrat, B., Gréchez-Cassiau, A., Guettier, C., Hastings, M. H. & Francis, L. (2004). Effects of chronic jet lag on tumor progression in mice. *Cancer Res., 64(21)*, 7879-85.

Mormont, M. C., Waterhouse, J., Bleuzen, P., Giacchetti S., Jami A., Bogdan A., Lellouch J., Misset JL., Touitou, Y. & Lévi, F. (2000). Marked 24-h rest/activity rhythms are associated with better quality of life, better response, and longer survival in patients with metastatic colorectal cancer and good performance status. *Clin Cancer Res.* 6(8),3038-45.

Rich, T., Innominato, P. F., Boerner, J., Mormont, M. C., Iacobelli, S., Baron, B., Jasmin, C. & Lévi, F. (2005). Elevated serum cytokines correlated with altered behavior, serum cortisol rhythm, and dampened 24-hour rest-activity patterns in patients with metastatic colorectal cancer. *Clin Cancer Res., 11(5)*, 1757-64.

Schernhammer, E. S., Laden, F., Speizer, F. E., Willett, W. C., Hunter D. J., Kawachi, I. & Colditz, G. A. (2001). Rotating night shifts and risk of breast cancer in women participating in the nurses' health study. *J Natl Cancer Inst, 93*, 1563-8.

Mhatre, M. C., Shah, P. N. & Juneja, H. S. (1984). Effect of varying photoperiods on mammary morphology, DNA synthesis, and hormone profile in female rats. *J Natl Cancer Inst, 72*, 1411-6.

Sahar, S. & Sassone-Corsi, P. (2007). Circadian clock and breast cancer: a molecular link. *Cell Cycle., 6(11)*, 1329-31.

Arjona, A. & Sarkar, D. K. (2005). Evidence supporting a circadian control of natural killer cell function. *Brain Behav Immun., 20(5)*, 469-76.

Mormont, M. C. & Levi, F. (1997). Circadian-system alterations during cancer processes: a review. *Int J Cancer, 70*, 241–7.

Vgontzas, A. N. & Chrousos, G. P. (2002). Sleep, the hypothalamic-pituitary-adrenal axis, and cytokines: multiple interactions and disturbances in sleep disorders. *Endocrinol Metab Clin North Am, 31*, 15-36.

Sansone, P., Storci, G., Tavolari, S., Guarnieri, T., Giovannini, C., Taffurelli, M., Ceccarelli, C., Santini, D., Paterini, P., Marcu, K. B., Chieco, P. & Bonafè, M. (2007). IL-6 triggers malignant features in mammospheres from human ductal breast carcinoma and normal mammary gland. *J Clin Invest., 117(12)*, 3988-4002.

Kramer, A., Yang, F. C., Snodgrass, P., Li, X., Scammell, T. E., Davis, F. C. & Weitz, C. J. (2001). Regulation of daily locomotor activity and sleep by hypothalamic EGF receptor signaling. *Science, 294*, 2511-5.

Hastings, M. H., Reddy, A. B. & Maywood, E. S. (2003). A clockwork web: circadian timing in brain and periphery, in health and disease. *Nat Rev Neurosci, 4*, 649-61.

Dantzer, R., O'Connor, J. C., Freund, G. G., Johnson, R. W. & Kelley, K. W. (2008). From inflammation to sickness and depression: when the immune system subjugates the brain. *Nat Rev Neurosci., 9(1)*, 46-56.

Henry, C. J., Huang, Y., Wynne, A., Hanke, M., Himler, J., Bailey, M. T., Sheridan, J. F. & Godbout, J. P. (2008). Minocycline attenuates lipopolysaccharide (LPS)-induced neuroinflammation, sickness behavior, and anhedonia. *J Neuroinflammation, 5*, 15.

Capuron, L., Ravaud, A. & Dantzer, R. (2000). Early depressive symptoms in cancer patients receiving interleukin 2 and/or interferon a-2b therapy. *J Clin Oncol, 18*, 2143–51.

Ohdo, S. (2007). Circadian Rhythms in the CNS and Peripheral Clock Disorders: Chronopharmacological Findings on Antitumor Drugs. *J Pharmacol Sci, 103*, 155-158.

Schibler, U. & Sassone-Corsi, P. (2002). A web of circadian pacemakers. *Cell, 111*, 919-22.

Lévi, F. (2006). The circadian timing system, a coordinator of life processes. implications for the rhythmic delivery of cancer therapeutics. *Conf Proc IEEE Eng Med Biol Soc.* Suppl: 6736-9.

Fu, L., Pelicano, H., Liu, J., Huang, P. & Lee, C. (2002). The circadian gene Period2 plays an important role in tumor suppression and DNA damage response in vivo. *Cell, 111*, 41-50.

Gery, S., Komatsu, N., Baldjyan, L., Yu, A., Koo, D. & Koeffler, H. P. (2006). The circadian gene per1 plays an important role in cell growth and DNA damage control in human cancer cells. *Mol Cell, 22*, 375- 82.

Lowrey, P. L., Shimomura, K., Antoch, M. P., Yamazaki, S., Zemenides P. D., Ralph, M. R., Menaker, M. & Takahashi, J. S. (2000). Positional syntenic cloning and functional characterization of the mammalian circadian mutation tau. *Science, 288*, 483-492.

Eide, E. J., Woolf, M. F., Kang, H., Woolf, P., Hurst, W., Camacho, F., Vielhaber, E. L., Giovanni, A. & Virshup, D. M. (2005). Control of mammalian circadian rhythm by CK1epsilon-regulated proteasome-mediated PER2 degradation. *Mol. Cell. Biol., 25*, 2795-2807.

Knippschild, U., Milne, D. M., Campbell, L. E., DeMaggio, A. J., Christenson, E., Hoekstra, M. F. & Meek, D. W. (1997). p53 is phosphorylated in vitro and in vivo by the delta and epsilon isoforms of casein kinase 1 and enhances the level of casein kinase 1 delta in response to topoisomerase-directed drugs. *Oncogene, 15*, 1727-1736.

Amit, S., Hatzubai, A., Birman, Y., Andersen, J. S., Ben-Shushan, E., Mann, M., Ben-Neriah, Y. & Alkalay, I. (2002). Axin-mediated CKI phosphorylation of beta-catenin at Ser a molecular switch for the Wnt pathway. *Genes Dev, 16*, 1066-1076.

Yang, W. S. & Stockwell, B. R. (2008). Inhibition of casein kinase 1-epsilon induces cancer-cell-selective, PERIOD2-dependent growth arrest. *Genome Biol., 9(6)*, R92.

Akten, B., Jauch, E., Genova, G. K., Kim, E. Y., Edery, I., Raabe, T. & Jackson, F. R. (2003). A role for CK2 in the Drosophila circadian oscillator. *Nat Neurosci., 6(3)*, 251-7.

Perea, S. E., Reyes, O., Puchades, Y., Mendoza, O., Vispo, N. S., Torrens, I., Santos, A., Silva, R., Acevedo, B., López, E., Falcón, V. & Alonso, D. F. (2004). Antitumor effect of a novel proapoptotic peptide that impairs the phosphorylation by the protein kinase 2 (casein kinase 2). *Cancer Res., 64(19)*, 7127-9.

Unsal-Kacmaz, K., Mullen, T. E., Kaufmann, W. K. & Sancar, A. (2005). Coupling of human circadian and cell cycles by the timeless protein. *Mol Cell Biol, 25*, 3109- 16.

Matsuo, T., Yamaguchi, S., Mitsui, S., Emi, A., Shimoda, F. & Okamura, H. (2003). Control mechanism of the circadian clock for timing of cell division in vivo. *Science, 302*, 255- 9.

Doi, M., Hirayama, J. & Sassone- Corsi P. Circadian regulator CLOCK is a histone acetyltransferase. *Cell, 125*, 497- 508.

Leone, M. J., Chiesa, J. J., Marpegan, L. & Golombek, D. (2007). A time to kill., and a time to heal. Pathophysiological interactions between the circadian and the immune systems *Physiological Mini-Reviews*, 2, 60-69.

In: Biological Clocks: Effects on Behavior... ISBN: 978-1-60741-251-9
Editors: O. Salvenmoser et al. pp. 139-161

Chapter 6

CHRONOECOLOGY OF NEOTROPICAL PRIMATES: THE SPIDER MONKEY ATELES GEOFFROYI

Jairo Muñoz-Delgado[1] and María Corsi-Cabrera[2]
[1]Chronoecology and Human Ethology Research-Group. Direction of Neurosciences. National Institute of Psychiatry Ramón de la Fuente Muñiz, Mexico City, and Faculty of Psychology, UNAM
[2]Sleep Laboratory. Faculty of Psychology, National Autonomous University of Mexico, UNAM

ABSTRACT

The emerging discipline denominated "chronoecology of behavior" is the result of the interaction between the work of behavioral ecologists –those who study behavioral phenomena in the natural environment– and chronobiologists, who explain biological rhythms on the basis of experimental studies carried out in the laboratory. Since the beginning of this century, this interaction has generated interest in understanding how biological rhythms behave in organisms living in their natural habitat. This approach requires field studies of rest/activity rhythms in order to obtain detailed hour/hour and day/day information about the possible effects of the modulation of certain environmental variables, such as photoperiod, temperature and the lunar sky, which may synchronize and/or mask behavioral activity. As a model for the study, a neotropical monkey species was chosen, the spider monkey, which is characterized by being widespread in America, from Mexico to southern Bolivia, and by its fusion-fission society and vast behavioral and vocal repertoire. In carrying out the studies, motor activity was registered using *Actiwatch* accelerometers during 24 continuous hours for 12 months, in combination with focal recordings of behavior and geophysical variables. On the basis of the results, the role of geophysical (astronomical and meteorological) variations as *zeitgeber* and/or the masking of the circadian rhythm, contributed valuable information for understanding alterations in physiology and behavior introduced by temporal adaptations to environmental changes. Finally, it is well-known that behavioral variations in non-human primates and *Homo sapiens* are influenced by astronomical and meteorological factors and by social, vocal, verbal and non-verbal conspecific interactions; therefore, it is essential to dissect these variations in order to study them and find out how the circadian rhythm modulates behavior and in what way social interaction influences the rhythms of circadian activity.

Keywords: geophysical variables, evolution, biological rhythms, behavior

INTRODUCTION

Chronobiology has studied the circadian patterns of temporal organization, physiological functions and the behavior of organisms, as well as the neural and genetic mechanisms that control them. In these studies, the patterns of rest/activity and sleep/wakefulness in several species have shown a fundamentally circadian organization. This evidence has been found primarily in rodents under controlled experimental conditions, but these studies have not taken into account the role of external variables in a natural environment, such as interaction with members of the same species and meteorological parameters like temperature, precipitation, humidity and the state of the sky, or astronomical parameters including sunrise and sunset, photoperiods, morning and evening twilight, and lunar, seasonal and annual phases.

Behavioral ecologists, meanwhile, have contributed knowledge about the behavioral habits of species in their natural environment, but have largely ignored biological rhythms [21, 24, Erkert, *pers. comm.*]. According to these studies, some species are primarily nocturnal, which means that they are active during the night and at rest during the day, while others are diurnal, but there are also the so-called twilight-active species, whose activity peak occurs during the morning and/or evening twilight. Moreover, some species –called cathemeral– present a poliphasic rhythm with multiple activity peaks during a 24-hour period. All of these behavioral patterns seem to share a common set of physiological mechanisms localized in the suprachiasmatic nucleus.

From an evolutionary point of view, the mechanisms that turn a nocturnal pattern into a diurnal, twilight-active or cathemeral circadian one are still unknown [10]. One possible explanation is that the same circadian cycle persists, but is "masked" by environmental factors such as temperature changes, the state of the sky and certain natural phenomena. This means that in the case of nocturnal species, the basic cycle is not inverted but that the absence of an activity peak is inhibited in the darkness and that the peaks present during daylight hours mask the resting-phase due to the activity imposed by factors such as the availability of food, protection from predators, noises emitted by members of the same and other species, etc. Another interpretation might be based on the direct effects of factors proper to the circadian cycle itself [10].

Chronoecology, a scientific field based on concepts that enable it to study circadian rhythms in natural conditions, unites the fields of chronobiology and behavioral ecology, and will thus contribute to improving our understanding of the natural factors that have an impact on endogenous rhythms [6, 21, 24, 28]. In this niche, experimental research and investigations in semi-natural and natural captivity with different types of monkeys has found its place, and now offers the possibility to inquire into how astronomical and meteorological factors or behavioral intra- and interspecific interaction affect both the circadian rhythms of rest/activity and sleep/wakefulness and neurophysiological activity.

This chapter contains a summary of the most important results of studies carried out with various monkey species and, especially, with a neotropical species, the spider monkey (*A. geoffroyi*) that, because it is diurnal, makes it possible to study the effects of the aforementioned parameters and to ascertain whether the behavioral patterns qualified as affiliative, directed towards themselves/oneself and aggressive show temporal variability.

Circadian Rhythms

The endogenous temporal organization of organisms' physiological and behavioral functions has an impact on the optimization of available resources and energy consumption. The temporal system that regulates the synchronization of an organism's internal processes with daily environmental events is called the circadian system, while the environmental factors that have an impact on it are called *zeitgeber*; these, in turn, synchronize the systems [2, 20, 27]. The most pronounced effect of the environmental *zeitgeber* is the natural cycle of light and darkness, but there are other physical and/or social variables that can reduce or increase the rest/activity rhythms of one or more physiological and/or behavioral variables, and thus lead to a daily or circadian rhythm that is masked to a greater or lesser extent [3, 13, 14, 15, 29, 33, 39].

In primates living in controlled laboratory conditions the effects of the light-darkness cycle, temperature changes, humidity, food availability, and even social variables, have been studied in the *Aotus Iemurinus* [8, 9, 12] and in the *Callithrix j. Jacchus* [8, 9], the *Eulemur fulvus* [8] and the *Saimiri sciureus* [4, 18].

However, the effect of geophysical phenomena such as sunrise and sunset, twilight periods, the hours of sunlight during the day, the lunar cycle, temperature, humidity and the state of the sky and climate have not been taken into consideration [17, 32]. Although it is known that certain social factors, such as visual and vocal interaction among subjects [15, 35], also have an effect on the circadian behavioral program that regulates the phase and period of the circadian clock [26], recordings and observation of behavior have been limited to the direct effects of certain environmental factors, such as temperature, in the *Eulemur fulvus collaris* [7] and the *Aotus azarai* [16].

A few studies of the relationship between the endogenous temporal system and the geophysical factors of rest/activity rhythms in primates have been undertaken in recent years by some chronobiologists and behavioral ecologists [5, 13, 16, 17, 22, 32, 33, 36,], all of whom have presented important data. It was found, for example, that the *Aotus azarai* is active during five hours in the day and 4 hours at night and that the duration of diurnal activity remains constant throughout the year despite seasonal changes between dry and rainy periods; nor does food availability change their activity. With an increasing time of moonlight, activity increases and the temperature of the environment predicts the activity during full-moon periods. Interaction among environment, temperature and moonlight are deciding factors in the patterns of the observed activities [16]. Contributions leading in this same direction were made in several studies carried out recently on cathemerality, which means a poliphasic rhythm with several rest-activity cycles during a 24 hour period, lunar periodicity and activity rhythms. Motor activity was recorded during one year using the actimetric method in the *Aotus azarai* by Fernández-Duque and Erkert [17]; they found increased activity during full-moon nights with a notable reduction during the day and some variations associated with changes of temperature, together with a cathemeral activity characterized by increased activity towards the twilight periods. The *Eulemur fulvus rufus,* originally from Madagascar, also presents cathemeral activity; diurnal activity increases with longer days, while nocturnal activity increases with longer nights in the cold season, and moonlight has an effect on the distribution of activities during the 24-hour cycle: more nocturnal activity takes place during the nighttime hours when the moon shines more brightly [22].

In order to find out if activity rhythms are regulated endogenously by the circadian time system which obeys the general rules for other mammals, or if it presents different characteristics in cathemeral species, Erkert and Cramer [11] carried out an experiment with an *Eulemur Fulvus albifrons* and *A. a. boliviensis* in controlled laboratory conditions with changing light-darkness cycles and light intensity. They found that, from the chronobiological point of view, both species are considered to be active in darkness. With *Propithecus v. verreauxi*, they carried out an analysis in the dry and rainy seasons related to motor activity rhythms in which they applied the actimetric method. They ascertained that although this primate is diurnal it has a bimodal activity pattern with important changes of activity in time and quantity during the year, with the peculiarity that in the mating season activity is unimodal, and a minor effect of moonlight; all of which suggests that thermoregulation and energy-related benefits were factors of primary importance in the activity patterns of these primates [13].

Rest-Activity Cycle in an American Primate, the Ateles Geoffroyi

A. geoffroyi

The *A. geoffroyi,* an American spider monkey (Figure 1) found from Mexico to southern Bolivia. Its fusion-fission social organization and vast behavioral and vocal repertoire, offers a model to study the role of geophysical (astronomical and meteorological) variations as *zeitgeber* and/or masking effects on the circadian rhythm, and the interaction between social behavior and circadian activity in field studies.

Determination of the Circadian Activity Rhythm of the *A. geoffroyi*: A Laboratory Experiment

First, the circadian activity rhythm of the *Ateles* had to be determined, as no previous information existed on this topic, assuming that they are provided with a physiological circadian system similar as all other mammals.

Motor activity

In order to study motor activity in laboratory, natural and/or semi-natural conditions, actimeters are used which are able to count and memorize impulses in intervals of 15, 30 seconds, and 1, 2, 5, 15 minutes, and up to periods of months, depending on the research interests for which they are used. For chronobiological studies, it is recommended to register in intervals of 5 minutes, as this has been shown to be an effective unit of analysis [9, 13, 17, 30, 32, 33] that allows researchers to obtain information over long-term periods of days, weeks, months or even years. In order to guarantee failure-free operation and protect the actimeters from damage by water or biting by conspecifics, they are housed in small aluminum boxes (40 mm x 32 mm x 13 mm) attached to a collar of laminated stainless steel.

The monkey was placed in an light-light 24 hours condition (LL) during 40 days and then in a 12:12 hours light-dark natural condition (LD) with an LD condition of 5 days between the experiments. Figure 2 shows the double plot actigram of the results of the periodic analysis; the average values of the activity patterns in the LL and natural LD conditions are indicated by arrows. It is clear that the spider monkey has a circadian rhythm of close to 24 hours and that it is a diurnal species, which means that it is active during light periods and rests during times of darkness.

This *tau*[1] values for each subject that formed part of the experiment were 24.04 h and 23.91 h for males, and 23.75 h and 23.87 h for females under the LL condition; under LD condition *tau* values for males were 24.20 h and 24.12h and for females 24.08 h and 24.16 h. A larger dispersion is observed in the LL condition than in the LD condition. Also, the end of the activity occurred significantly earlier in the LL condition than during the LD cycle (Figure 2). Thus, we can conclude so far that the spider monkey presents periodic changes of activity in the LL and LD cycles indicating the presence of a pacemaker that has the intrinsic capacity to oscillate when the organism is in extreme conditions of continuous light, and to get synchronized by the photoperiod in LD conditions. An excessive quantity of light as in LL induces a great dispersion among individual oscillators.

[1] *tau*: circadian period

Figure 1. Neotropical primate. Spider monkey *A. geoffroyi* in one of the typical positions which seams to be synchronic. The spider monkey is an American anthropoid and its habitat reaches from the Mexican State of Veracruz to the north of Bolivia.

Rest-Activity Rhythm of the *A. geoffroyi* in Semi-Captivity

Astronomical variables

The recording of motor activity during 11 months in conditions of semi-captivity with exposure to the natural environment of LD ratifies that the *Atleles* is active during the day and rests at night (Figure 3a). The period of activity ended at an average of 1 h 10 min after sunset, thus showing a negative phase position. Two activity peaks were found, one in the morning hours and the other in the early afternoon (Figure 3b). The coefficients from multiple regressions showed significant correlations between the astronomical variables and the rest-activity parameters; monkeys activity time, and onset and end of activity were significantly correlated with sunrise and solar day duration, as follows: activity period was correlated with solar day duration ($t = -2.34$ $p = 0.023$) (Figure 4), with sunrise ($t = -2.38$ $p = 0.021$) and with minimum temperature ($t = 2.06$ $p = 0.044$); total activity was correlated with sunset ($t = -2.42$ $p = 0.019$) and twilight end ($t = 2.81$ $p = 0.007$); the time of the morning activity peak was correlated with solar day duration ($t = 2.2$ $p = 0.032$), sunrise ($t = 2.27$ $p = 0.027$) and twilight end ($t = -5.13$ $p = 0.043$).

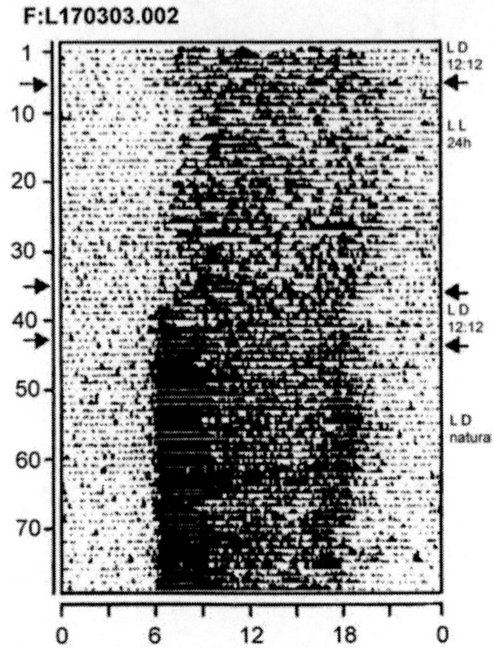

Figure 2. Single plot actogram of the rest-activity rhythm of a male spider monkey *A. geoffroyi* in the two experimental conditions to determine the circadian rhythm. Histogram lines show the 5-min original data of an uninterrupted recording period of 80 consecutive days. Line spaces correspond to 100 counts per 5-min bins. During the first five days the monkeys were kept in individual cages in a room under 12:12 light-darkness condition (LD) to habituate them to experimental conditions. In the following 30 days they were exposed to continuous light during 24 hours a day (LL). The food was given randomly at any day or night time with *ad libitum* water availability. Afterwards, they were kept again during 5 days in LD12/12 conditions before they were returned to the community cage where they stayed for the next 30 days with the rest of the group in natural LD conditions.

Meteorological variables

There were some randomly distributed sultry and cloudy days during the dry half of the year when it was possible to observe that climatic factors did have an effect on the monkeys' activity patterns. As can be seen in Figures 5 and 6, climatic factors significantly influenced the duration and times of onset and end of activity. Activity time was shorter on mild as compared to hot, sultry days (F=6.66; P<0.0003) and began earliest on sultry days (F=13.49; P<0.0001). Cloudiness affected only the end of activity, which ended earlier on cloudy days as compared to clear ones (F=5.09; p<0.009).

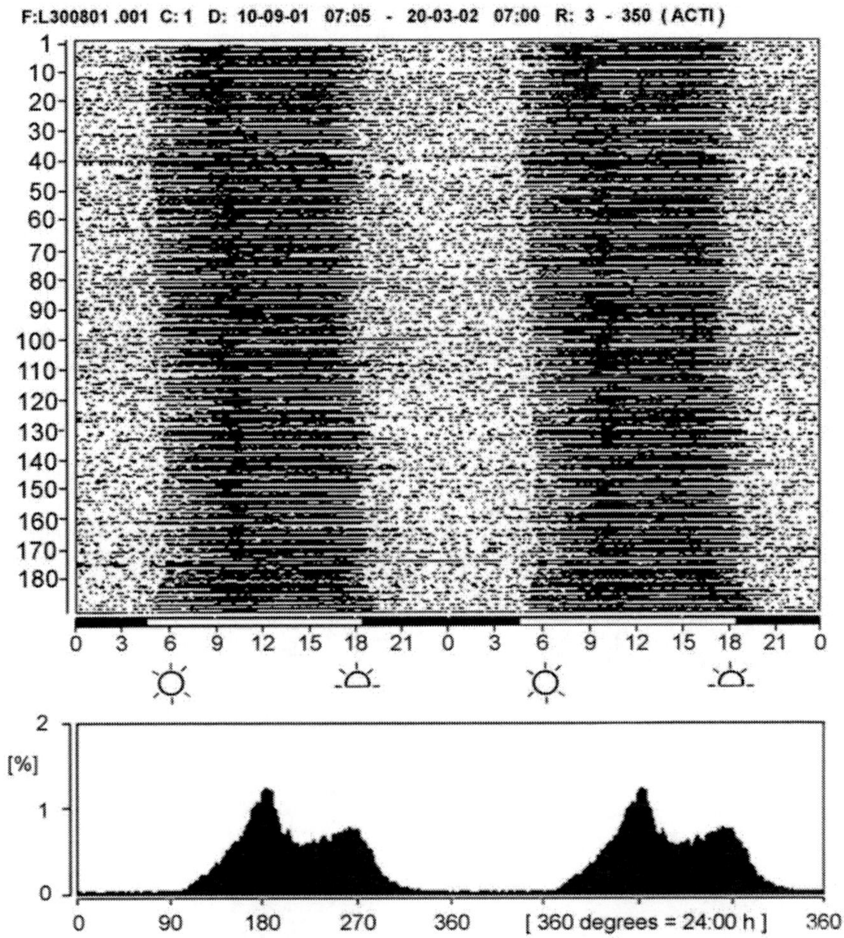

Figure 3. Upper diagram: Double plot actogram of the rest-activity rhythm in a male spider monkey *A. geoffroyi* living in exterior captivity in natural environmental conditions at the "Parque de la Flora y Fauna Silvestre Tropical Pipiapan" of the Veracruz University, Mexico. Histogram lines show the 5-min original data of an uninterrupted recording period of 190 consecutive days. Line spaces correspond to 100 counts per 5-min bins. Below: Double plot of mean activity of the same subject averaged over the whole recording period. Ordinate values are given as a percentage of the averaged total amount of activity per day.

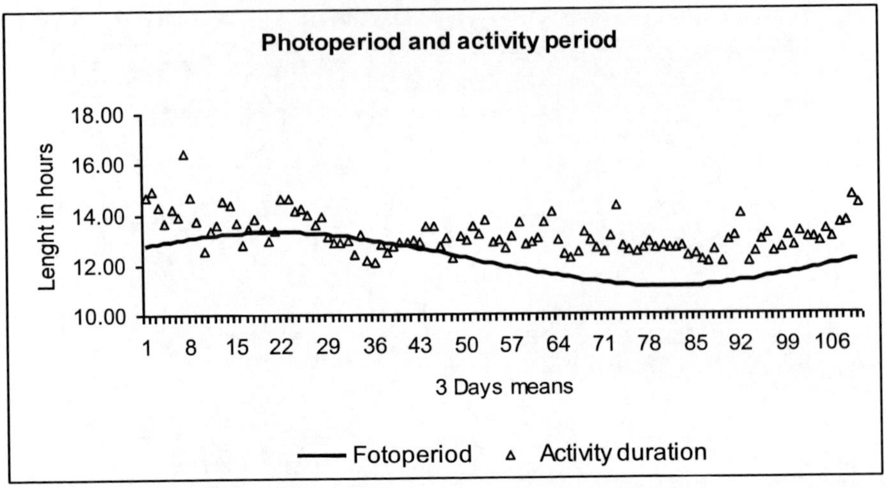

Figure 4. Time course of circadian rhythm during 11 months of continuous recording. Solar-day duration (photoperiod) and daily average of duration of activity time of five spider monkeys *A. geoffroyi* living in exterior captivity in natural environmental.

Our results clearly show a modulation of the rest-activity patterns as an effect of astronomic parameters, which work as *zeitgeber*, and a masking effect of the meteorological factors, while the artificial variables introduced by human interaction have only a weak effect [32].

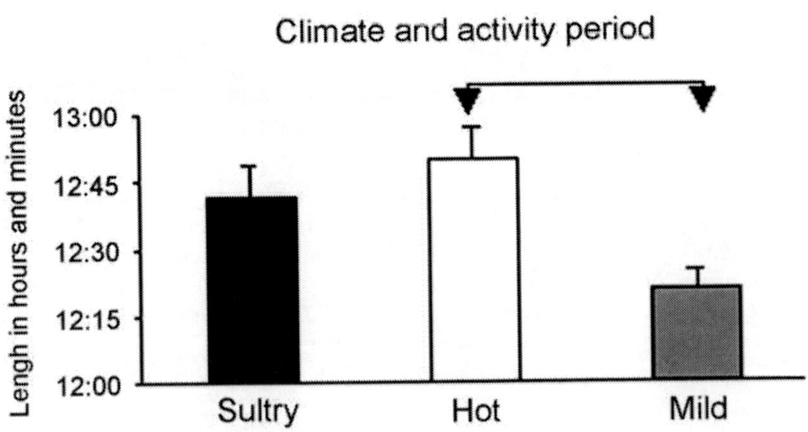

Figure 5. Shows the average and the standard error of the duration of the activity period of five spider monkeys *A. geoffroyi* living in exterior captivity in natural environmental conditions during sultry (12.7 ± 0.1), hot (12.83 ± 0.11) and mild (12.35 ± 0.06) days. The arrows indicate significant differences.

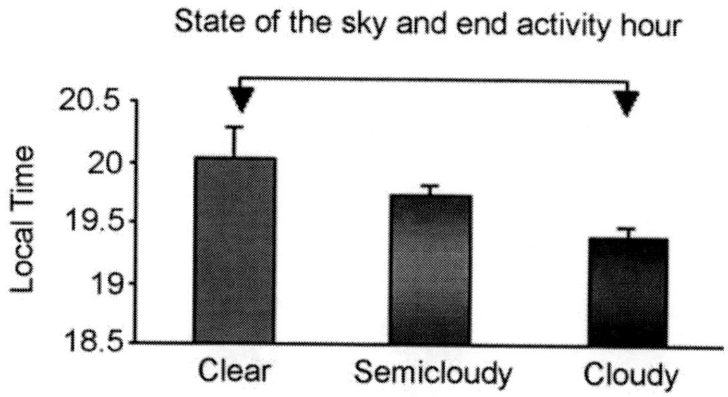

Figure 6. Average and the standard error of the time of day when the activity period ends in five spider monkeys *A. geoffroyi* living in exterior captivity in natural environmental conditions in clear (20.03 ± 0.26), semi clouded (19.74 ± 0.09) and clouded (19.4 ± 0.09) skies. The arrows indicate significant differences.

The analysis of 11 months of the motor activity of *A. geoffroyi* in conditions of semi-natural captivity, showed that activity increases during the rainy season while it is reduced during the dry season (Figure 7), a finding that might be related to room temperature. The dry season is the hottest period, with annual average temperatures of 27°C, so it is possible that activity diminishes in this period as a means of conserving energy. This indicates a transitory dependence of the motor activity on temperature which may be acting a masking factor (*unpublished results*).

Lunar cycles

The lunar cycle also has an influence on motor activity. We found gender and group differences in the amount of activity during the three hours after evening twilight and at daybreak during full-moon nights in a 6-month analysis (Figure 8). The *Atleles* tends to show less activity during the new moon and to initiate its activity earlier at full-moon. Activity increases at full-moon from 19:00 to 07:00 o'clock, except for the period from 01:00 to 04:00 o'clock. The absence of differences between 01:00 and 04:00 o'clock could be due to the fact that the sleep tendency is highest during this time, with the concomitant low body temperature and strong impulse towards uncontrollable sleep, as has been demonstrated in other species. Afterwards, in the early morning hours from 07:00 to 10:00 am, an inactive period is observed that might be a consequence of the lack of sleep due to alterations in the resting period during nights with bright moonlight, but this factor has not yet been studied [32]. The modification of nocturnal activity during moonlit nights indicates the plasticity of the circadian systems and implies a visual system with a good sensitivity that allows adaptations to the daily variations in light intensity. From the evolutionary point of view, however, the internal nature of lunar rhythms is still unknown [31].

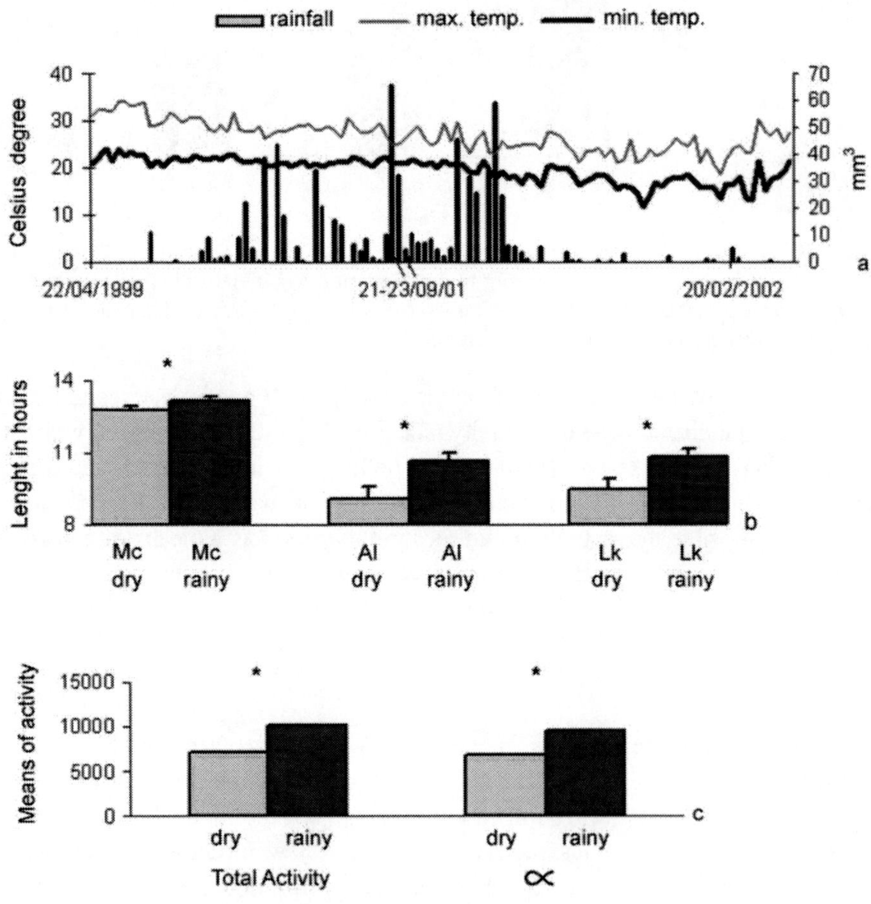

Figure 7. Distribution of meteorological variables; rainfall, maximum temperature and minimum temperature in 11 month of recording in the "Parque de la Fauna y la Flora Silvestre Tropical Pipiapan", Catemaco, Veracruz, Mexico (a). Average and standard error of the duration of the activity period in the dry and rainy seasons in spider monkeys, *A. geoffroyi,* living in exterior captivity in natural environmental conditions (b). Average and standard error of total activity in the dry and rainy seasons of the same monkeys (c).

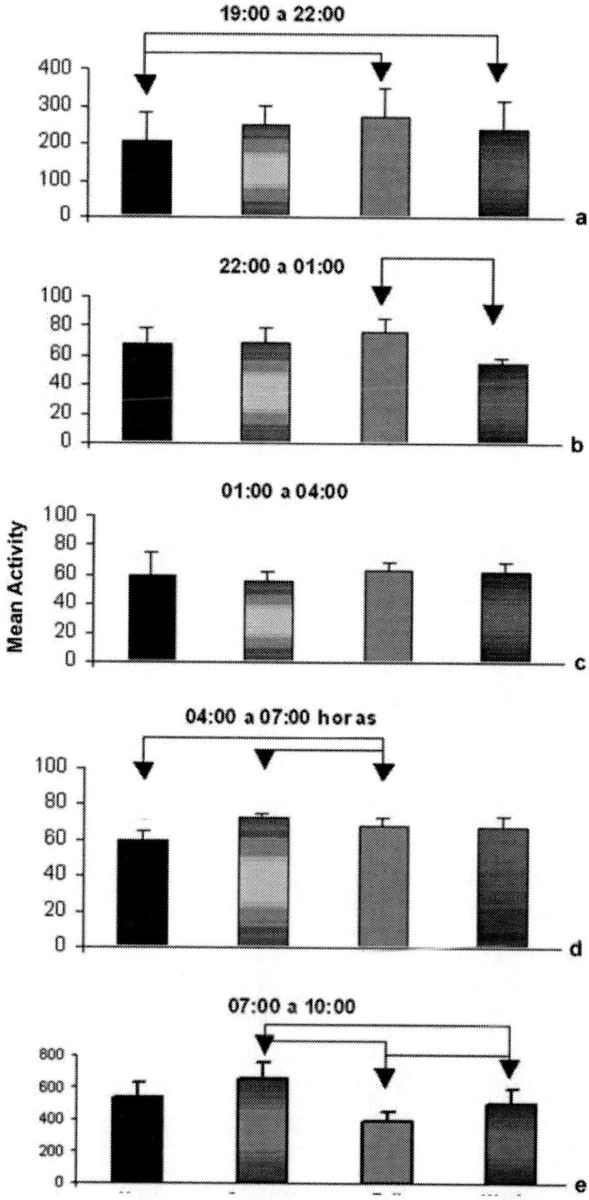

Figure 8. Average and standard error of the total of motor activity of five spider monkeys, *A. geoffroyi,* living in exterior captivity in natural environmental conditions in each lunar phase during 6 consecutive months at different times of day: 19:00 to 22:00 (a), 22:00 to 01:00 (b), 01:00 to 04:00 (c), 04:00 to 07:00 (d), and 07:00 to 10:00 (e). Arrows indicate significant differences between moon phases.

Rest-Activity Rhythm of the *A. geoffroyi* in Out-Door Semi-Captivity

Change of living conditions from outdoor captivity to an electrified fence enclosure in the natural environment: impact on the circadian rest-activity rhythm.

An interesting phenomenon we were able to observe in our studies was that the change in living conditions from outdoor captivity (Figure 9a) to an electrified fence enclosure (Figure 9b) has an impact on the rest-activity rhythm and on behavioral and physiological activities. Upon examining Figure 10, we can see that the circadian rest-activity rhythm is organized and responds to the *zeitgeber* in the condition of outdoor captivity, in which the monkeys were kept for almost two years; we can also observe changes in the activity peaks in the periodograms caused by the different living conditions (Figure 10a and b). Later, on day 30, the group was moved to an electrified fence enclosure, and a disorganization of the rest-activity rhythm can be observed in the thirty days that followed (Figure 10a and c). Then we can see the circadian adaptation to the new environment (Figure 10 a and d). The mean values of the alpha activity were also different for each subject, 91.85 in one monkey and 42.46 in the other monkey in the outdoor captivity condition. When the animals were moved to the electric enclosure the mean of the alpha activity in the following 30 days of adaptation was 82.5 and 21 respectively. In the following 30 days, period in which the animals are adapted to the electric enclosure, the mean of the alpha activity for each subject was 116.4 and 42.46 respectively (*Unpublished results*).

A characteristic of most of the species –in this case primates– is a bimodal activity pattern with two activity peaks during the day, or at morning and evening twilight. In the *Ateles*, however, in experimental laboratory conditions, in conditions of outdoor captivity and of semi-natural captivity, we have found bimodal activity with differences in the value and duration of the first activity peak. The first peak in outdoor captivity and semi-natural conditions shows a significantly long duration and a high level compared to the second peak, while in experimental laboratory conditions the two peaks occupy almost the same level both in amplitude and duration [33]. The differences in the amplitude and duration of the two activity peaks in the different conditions might be related to the availability of food. In the outdoor captivity condition the first peak showed an activity mean of 2.34 and 1.68, and the second one a mean of 0.39 and 0.63 respectively (Figure 10b). During the adaptation period in the electric enclosure, the first peak showed an activity mean of 0.95 and 1.01, while the second peak showed an activity mean of 0.78 and 0.78 respectively (Figure 10c). Finally, when the animals had already adapted to the electric enclosure condition, the mean of the activity of the first peak was of 1.52 and 1.68, and the second peak was of 2.33 and 2.11 respectively (Figure 10d) (*Unpublished results*).

Figure 9. The exterior captivity in the natural environment where monkeys were kept originally at the "Parque de la Flora y Fauna Silvestre Tropical Pipiapan" of the Veracruz University, Mexico is shown in a. It is a construction with rectangular meshes measuring 12 m in length, 6 m in width and 3 m in height. The middle of the enclosure is covered by a laminated roof which protects the monkeys from the rain. The electric fence enclosure in the same natural environment is shown in b. It has a diameter of 250 m, equivalent to ¼ hectare where monkeys are free to forage and climb up to the highest parts of the trees.

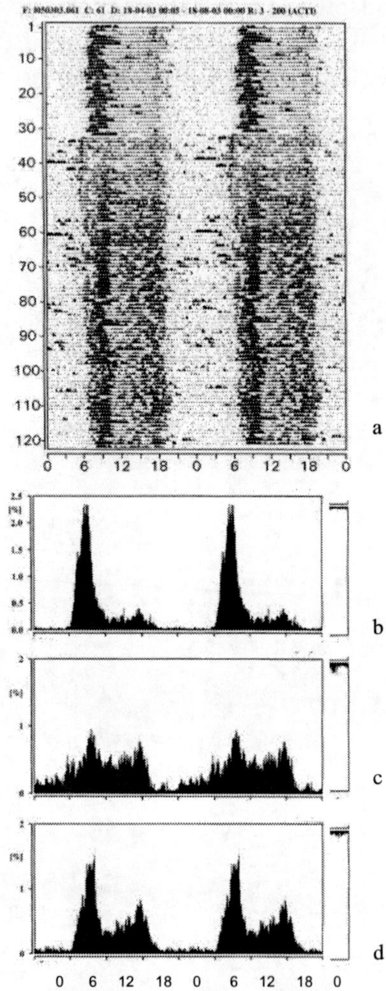

Figure 10. Double plot actogram (a) of one spider monkeys *A. geoffroyi* of the last 32 days of recording during their stay in exterior captivity in natural environmental conditions and the first 60 days in an electric fence enclosure in the same natural environment; on day 32 they were moved to the electric fence enclosure. From day 32 to day 60 a desynchronization of the circadian rest-activity rhythm is observable. Between days 62 and 122 it can be observed how the rest-activity cycle is adapted to the new living conditions. Figures b, c and d show double plots of average activity during exterior captivity (b), during the first 30 days (b) and the last 30 days (c) in the electric fence enclosure. We can observe how the average activity during 24 hours is organized in b, disorganized in c, and reorganized in d. The morning peak is more pronounced than the afternoon peak in b, whereas the afternoon peak is more pronounced in d.

Behavioral categories with daily temporal variation

In addition to the scarce information available on behavior associated with the rest-activity rhythm, another area that remains to be explored is precisely what behaviors are associated with geophysical variations and how they are influenced by those factors.

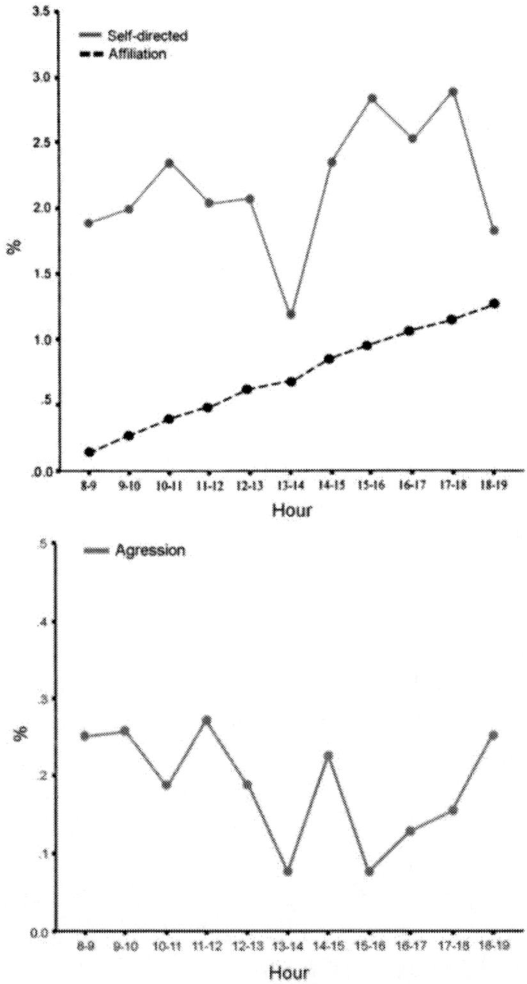

Figure 11. Percentage of aggression, affiliation and self-directed behaviors at different times of day of 4 spider monkeys *A. geoffroyi* in electric fence enclosure in natural environmental conditions during 90 days.

Behavior

The recording of behavior with a video camera with the images being saved on magnetic tapes, including both the quantity and quality of motor activity, can be focal, which means a continuous monitoring of one or two individuals during a previously determined period of time, i.e. during sunlight hours or twilight periods, or one or various behaviors on all occasions [25, 36]. With these recordings it is possible to obtain behavioral information in real time and thus establish correlations between periods of motor activity or specific behaviors and the astronomical and meteorological parameters under study. Video recordings have proven to be an efficient means of obtaining data on specific behaviors, for example during sleep [1, 34, 37, 38]. Additionally, vocalizations and environmental sounds can be recorded on audio tapes during the same periods [19].

Table 1. Behavior catalog.

Agression	Agression	Affilation	Self-directed	Submission
mild	*rough*			
displace	push	lie next to	grooming	flee
stand up straight	fight	sit next to	brachiate	scream
bare teeth	shake	embrace	hang	avoid
open mouth threat	kick	ask for grooming	swing	
charge	pull hair	follow	eat	
fixed stare	slap	support	forage	
chase	seize food	huddle	rest	
branch shaking		grooming		

We carried out a sequence of behavioral observations during the day, described in Table 1, in order to determine if the behavioral expression and the rest-activity rhythm show temporal rhythmic variations, and found that there was a significant increase in affiliation behaviors along the day ($F = 10.89; p = 0.00$), while self-directed behaviors increased between 15:00 and 16:00, and between 17:00 and 18:00 ($F = 11.78; p = 0.00$). Aggressive behaviors showed a slight increase in the early morning hours between 08:00 and 10:00 and between 18:00 and 19:00 ($F = 2.24; p = 0.013$). Finally, submissive behaviors did not show a tendency to occur more frequently at any specific time of day ($F = 0.68; p = 0.73$) (see Figure 11) [23] *(unpublished results).*

CONCLUSION

These observations are congruent with the findings of ecological studies, thus proving that the social organization of Ateles is of a fusion-fission nature. They also demonstrate that the spider monkey has an internal clock that is expressed both in motor activity and in behavioral activity, and also presents a temporal organization that suggests a strategy for optimizing resources and energy conservation and for reducing intra-group rivalries. The importance of carrying out studies in natural conditions and/or semi-natural and laboratory conditions rests upon the possibility of obtaining information about the adaptation processes of circadian clocks to changes and variations in the artificial environment as well as to variations in the natural environment. These differences enable us to explain the role that these parameters play as *zeitgeber* and/or masking factors of the circadian rhythm and its adaptive value. Research on these phenomena in controlled laboratory conditions and in semi-natural and natural conditions, constitutes a bridge between chronobiological studies and research in the field of behavioral ecology [21], which converge into one of the recent subdisciplines of chronobiology, called behavioral chronoecology. It is important to explain the interaction between organisms' internal temporal mechanisms and environmental changes in order to understand the alterations in physiology and behavior introduced by temporal adaptations and environmental change [24].

If it is true that behavioral variations in non-human primates and *Homo sapiens sapiens* are influenced by astronomical or meteorological factors, by verbal and non-verbal social interaction and by economic-ethnographic relations such as beliefs, usages and customs, then it is important to dissect these variations in order to find out how they affect circadian activity rhythms. Finally, studies of the rhythms of behaviors categorized as social, aggressive, etc., provide important information related to activity peaks and can contribute to determining a species' adaptation processes from the evolutionary point of view.

THE FUTURE OF CHRONOECOLOGICAL RESEARCH

Based on the schemes used for research on rhythms in semi-natural and natural conditions, it was possible to explain the presence of circadian rhythms in diurnal and nocturnal species. But with advances in techniques of solid recording, such as the actimetric method, the existence of other rhythms, previously unknown, in species that have been considered diurnal or nocturnal, has become evident. Today we speak of cathemeral primates, as well as of rhythms modulated by the lunar phases in the nocturnal monkey *Aotus*, which belongs to the subspecies *Azarai*, and in the diurnal *Ateles*. These results will provide valuable information that could help to explain the transition from nocturnal to diurnal species, and to isolate the environmental phenomena that forced some species to change to diurnal activity [10], as is expressed in the human species. This is why it is important to carry out this type of study with as many species of primates as possible: the diurnal or nocturnal nature of a species has simply been assumed on the basis of studies carried out during the day, but those analyses ignore the fact that an organism's life takes place during periods of 24 hours. Discovering a species' circadian rhythmicity in the context of its socioecology represents a factor that can help to optimize research in any field of science.

Acknowledgments

We thank Master Silke Trienke for the English version and Paul Kersey PhD for the revised version of this manuscript and Raúl Cardoso for his work on the figures. Research by the present authors was funded in part by the Instituto Nacional de Psiquiatría Ramón de la Fuente Muñiz project 3330, and the Zoology Institute of Tuebingen University for lending the Actiwatch equipment.

References

Adame, M; Cerda, A; Zubirán, P; Páez, F; Santillán-Doherty A; Mondragón-Ceballos R. Confiabilidad entre observadores no experimentados de conductas de reposo-actividad en macacos cola de muñón *Macaca arctoides*, en cautiverio exterior. *Salud Mental.*, 2000, 23(2), 16-20.

Aschoff, J. Exogenous and Endogenous Components in Circadian Rhythms. *Cold Spring Harb Symp Quant Biol*, 1960. 25, 11-28.

Aschoff, J. Biological Rhythms, Handbook of Behavioral Neurobiology, vol 4, New York: *Plenum Press*, 1981.

Aschoff, J; Tokura, H. Circadian Activity Rhythms in Squirrel Monkeys: Entrainment by Temperature Cycles. *Journal of Biological Rhythms*, 1986, 1, 91-99.

Cheyne, S. Effects of Meteorology, Astronomical Variables, Location and Human Disturbance on the Singing Apes: *Hylobates albibarbis*. *American Journal of*

Primatology., 2008, 70(4), 386-392.

DeCoursey, P. Diversity of Function of SCN Pacemakers in Behavior and Ecology of three Species of Sciurid Rodents. *Biological Rhythm Research*, 2004, 35(1/2), 13-33.

Donati, G; Borgognini, S. The Cathemeral Activity Patterns of the Collared Brown Lemur, *Eulemur fulvus collaris*: an Account from the Littoral Forest (Abstract). Abstract of the XIXth *Congress of the International Primatological Society*., 2002, Beijing, China.

Erkert, HG. Lighting Requirements of Nocturnal Primates in Captivity: a Chronobiological Approach. *Zoo Biology*, 1989, 8, 179-191.

Erkert, HG. Chronobiological Aspects of Primate Research, In: Setchell, J; Curtis, J. *Field and Laboratory Methods in Primatology*. United Kingdom: Cambridge University; 2003, 252-270.

Erkert, HG. Diurnality and Nocturnality in Nonhuman Primates: Comparative Chronobiological Studies in Laboratory and Nature. *Biological Rhythm Research*, 2008, 39(3), 229-267.

Erkert, HG; Cramer, B. Chronobiological Background to Cathemerality: Circadian Rhythms in *Eulemur fulvus albifrons* (Prosimii) and *Aotus azarai boliviensis* (Anthropoidea). *Folia Primatologica*., 2006,77, 87-103.

Erkert, HG; Gröber, J. Direct Modulation of Activity and Body Temperature of Owl Monkeys (*Aotus lemurinus griseimembra*) by Low Light Intensities. *Folia Primatologica*., 1986, 47, 171-188.

Erkert, HG; Kappeler, PM. Arrived in the Light: Diet and Seasonal Activity Patterns in Wild Verreaux's sifakas (*Propithecus v. Varreauxi*, Primates: Indriidae). *Behavioral Ecology and Sociobiology*. 2004, 57, 174-186.

Erkert, HG; Nagel, B; Stephani, I. Light and Social Effects on the Free-running Circadian Activity Rhythm in Common Marmosets (*Callithrix jacchus*; Primates: Social Masking, Pseudospliting, and Relative Coordination). *Behavioral Ecology and Sociobiology*, 1986. 18, 443-452.

Erkert, HG; Schardt, U. Social Entrainment of Circadian Activity Rhythms in Common Marmosets, *Callithrix j. jacchus* (Primates). *Ethology*, 1991. 87, 189-202.

Fernández-Duque, E. Influences of Moonlight, Ambient Temperature, and Food Availability on the Diurnal and Nocturnal Activity of Owl Monkeys (*Aotus azarai*). *Behavioral Ecology and Sociobiology*, 2003, 54, 431-440.

Fernández-Duque, E; Erkert, HG. Cathemerality and Lunar Periodicity of Activity Rhythms in Owl Monkeys of the Argentinean Chaco. *Folia Primatologica*, 2006, 77, 123-138.

Fuller, CA; Sulzman, FM; Moore-Ede, MC. Circadian Control of Thermoregulation in the Squirrel Monkey *Saimiri Sciureus*. *American Journal of Physiology*, 1979, 236(3), 153-161.

Geissman, Th. Tape Recording Primate Vocalizations. In: Setchell, J; Curtis, J. *Field and Laboratory Methods in Primatology*. United Kingdom: Cambridge University; 2003, 228-238.

Gruart, A; Delgado JM; Escobar C; Aguilar Roblero R. Los relojes que gobiernan la vida. México: Fondo de Cultura Económica; 2002.

Halle, S; Stenseth, NC. Activity Patterns in Small Mammals. Berlin: Springer; 2000.

Kappeler, PM; Erkert, HG. On the Move Around the Clock: Correlates and Determinants of Cathemeral Activity in Wild Red-Fronted Lemurs (*Eulemur fulvus rufus*). *Behavioral Ecology and Sociobiology*, 2003. 54(4), 359-369.

Landeros, T. Variación diaria del comportamiento agresivo en un grupo de monos araña (*Ateles geoffroyi*). *Tesis de Licenciatura*. Antropología Física. ENAH, Mexico, 2008.

Marques, MD; Waterhouse, J. Rhythms and Ecology-do Chronobiologists Still Remember Nature? *Biological Rhythm Research*, 2004, 35(1/2), 1-2.

Martín, P; Bateson, P. *La medición del comportamiento*. Madrid: Alianza Editorial; 1986.

Mistlberger, R; Skene, D. Social Influences on Mammalian Circadian Rhythms: Animal and Human Studies. *Biological Review*, 2004, 79, 533-556.

Moore-Ede, MC; Sulzman, FM; Fuller, Ch. The Clocks that Time Us. Physiology of the Circadian Timing System. USA: Harvard University Press; 1982.

Morgan, E. Ecological Significance of Biological Clocks. *Biological Rhythm Research*, 2004, 35(1/2), 3-12.

Mrosovsky, N. Masking: History, Definitions, and Measurement. *Chronobiology International*, 1999, 16(4), 415-429.

Muñoz-Delgado, J. Variaciones del período de reposo-actividad del mono araña *Ateles geoffroyi*, relacionados con cambios diarios, lunares y anuales. *Tesis de doctorado*. Facultad de Psicología, UNAM, México, 2004.

Muñoz-Delgado, J. A propósito de los ritmos de la luna. *Ciencia*, 2008, 59(1), 48-56.

Muñoz-Delgado, J; Corsi-Cabrera, M; Canales-Espinosa, D; Santillán-Doherty, AM; Erkert, HG. Astronomical and Meteorological Parameters and Rest-Activity Rhythm in the Spider Monkey *Ateles geoffroyi*. *Physiology and Behavior*, 2004, 83(1), 107-117.

Muñoz-Delgado, J; Fuentes-Pardo, B; Euler Baum, A; Lanzagorta, N; Arenas-Rosas,

R; Santillán-Doherty, AM; Guevara, MA; Corsi-Cabrera, M. Presence of a Circadian Rhythm in the Spider Monkey's *Ateles geoffroyi* Motor Activity. *Biological Rhythm Research*, 2005, 36(1/2), 115-121.

Muñoz-Delgado, J; Luna-Villegas, G; Mondragón-Ceballos, R; Fernández-Guardiola, A. Behavioral Characterization of Sleep in Stumptail Macaques (*Macaca arctoides*) in Exterior Captivity by Means of High-Sensitivity Videorecording. *American Journal of Primatology*, 1995, 36, 245-249.

Pálkova, M; Sigmund, L; Erkert, HG. Effect of Ambient Temperature on the Circadian Activity Rhythm in Common Marmosets, *Callithrix j. jacchus* (Primates). *Chronobiology International*. 1999,16, 149-161.

Papailious, A; Sulliven, E; Cameron, JL. Behaviors in Rhesus Monkeys (*Macaca mulatta*) Associated with Activity Counts Measured by Accelerometer. *American Journal of Primatology*., 2008, 70(2), 185-190.

Rowe, N; Myers M. Photography and Video for Field, Researchers. In: Setchell, J; Curtis, J. *Field and Laboratory Methods in Primatology*. United Kingdom: Cambridge University; 2003, 239-251.

Santillán-Doherty, AM; Muñoz-Delgado, J; Arenas, R; Márquez, A; Cortés, J. Reliability of a Method to Measure Novelty-Seeking in Nonhuman Primates. *American Journal of Primatology*, 2006, 68, 1098-1113.

Sulzman, FM; Fuller, CA, Moore-Ede, MC. Environmental Synchronizers of Squirrel Monkey Circadian Rhythms. *Journal of Applied Physiology*, 1977, 43, 795-800.

In: Biological Clocks: Effects on Behavior… ISBN: 978-1-60741-251-9
Editors: O. Salvenmoser et al. pp. 163-176

Chapter 7

ON CLOCKS, CHAOS AND CANCER: A BIODYNAMIC APPROACH TO CANCER

Federico Cardona*

Wiedemann Parkklinik, Biomedical Research Dept., Markdorf, Germany

ABSTRACT

Self-organizing of open dynamic systems far from thermodynamic equilibrium gives rise to complexity in space and time, thus creating new structures including biological clocks. In fact, periodic oscillations can be used to control deterministic chaos. The human organism is a multioscillator, which is governed by the master clock in the hypothalamus. We have previously observed a periodic oscillation of free radical production that functions synchronously to the master clock and is apparently generated by it. We found as well that this oscillation, possibly mediated by mitochondrial activity, ceases to function in humans suffering from advanced cancer diseases. Alterations of mitochondrial bioenergetics have been observed to be a general feature of malignant cells and could progressively diminish the distance from thermodynamic equilibrium necessary to maintain multicellular complexity. Self-organization could thus break down, leading to malignant development as a survival strategy of tumor cells.

Keywords: Bioenergetics, Biological clocks, Cancer, Circadian Rhythm, Lipid Peroxidation, Mitochondria, Nonlinear Dynamics.

* Corresponding Author: fcardona@web.de

Introduction

The second law of thermodynamics states that the universe constantly aims towards a state of maximal entropy, a measure of its disorder. A coffee cup can easily be broken, but there is little probability for the pieces to spontaneously recover their original form again. A cup of coffee will get cold within minutes but quite improbably will get spontaneously hot again. In both cases, a shorter distance to thermodynamic equilibrium has been reached, implying more disorder. These simple examples are not trivial but represent rather the basis of life.

Many parts of the world around us, particularly living nature, display an increasing order the more they are able to attain a condition situated far from that thermodynamic equilibrium, which apparently contradicts classic thermodynamics. Living organisms may thus develop an increasingly high degree of complexity.

Self-Organization: An Overview

Life on earth fundamentally arises from sunlight hitting our planet, which is used by nature to build up conditions fundamental for life and its environment. The functioning of living systems is based on a continual input of energy and matter, as well as an output of heat, which they constantly discharge ("dissipate") into their environment giving rise to complexity. They are therefore referred to as "open systems".

The "history" of a system increasing its distance from thermodynamic equilibrium and a higher complexity degree is illustrated by the "bifurcation diagram" (Figure 1). It shows the distance from equilibrium of each single point of a system.

Previous to and after the dark zones of the diagram, bright recurrent phases, so called "laminar phases" of stable cyclic behavior (bifurcations), appear (Figure 2). The alternation of chaotic and periodic behavior is referred to as "intermittency", which can be considered to be the memory of nonlinear systems to remind them of their initial nonchaotic (periodic) attractors [22]. Therefore, clocks are situated in the laminar phases of a system's bifurcation diagram as part of that memory and reflect the present or past of a given system, e.g. the world around us. Within the dark chaotic zones, complex systems spontaneously acquire the ability to spontaneously create complexity and an infinite number of unpredictable spatiotemporal structures.

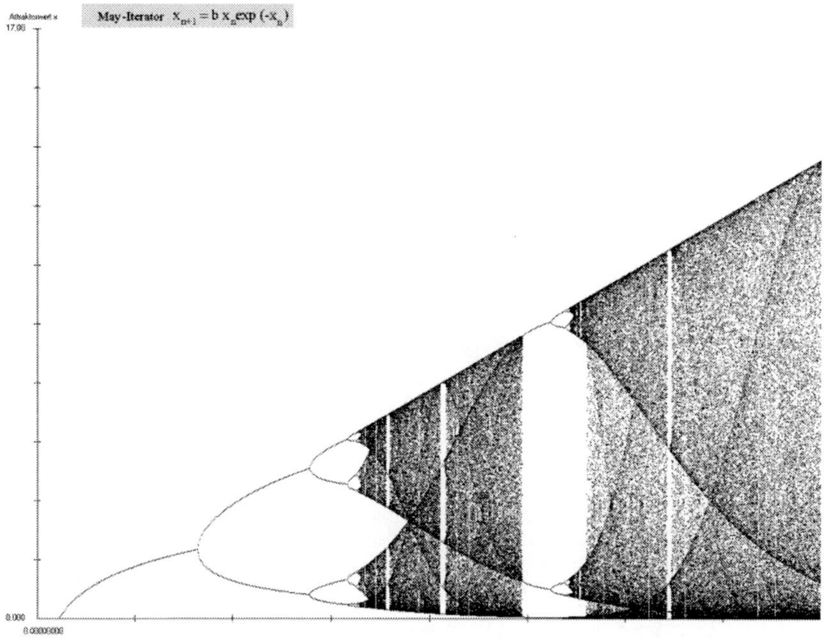

Figure 1. Bifurcation Diagram.

Much of our present knowledge in this area is due to the work of the Russian-Belgian physicist Ilya Prigogine, Nobel award winner in 1977 [25]. Prigogine performed profound research describing the above mentioned time dependent "open systems", which he termed "dissipative structures".

In self-organizing systems, pattern formation occurs through interactions internal to the system, without intervention by external directing influences [4]. As a result of interactions, new properties emerge abruptly in such systems, which represent more than the sum of their parts and are thus nonlinear. Examples of self-organizing units that form an interactive network with new collective properties are individual units e.g. 10^5 genes to build 300 cell types as an adaptive interactive network [8]. The nonlinear nature of living processes turns out to be crucial for understanding how biological systems interact with the external environment [29].

The dynamical states that result from self-organizing processes may have features such as excitability, periodicity, bistability, chaos or spatiotemporal pattern formation and all of these can be observed in biological systems [29]. Nonlinear dynamics in self-organizational processes provide therefore a thoroughly new approach to understand complexity in living systems. All these spatially and temporally ordered structures indicate the existence of mechanisms of self-organization acting within the cell, which cannot be deduced from genetic and molecular biological knowledge alone [20]. One of the revelations of self-organization studies is that the richness of structures observed in nature does not require a comparable richness in the genome but can arise from the repeated application of simple rules by large numbers of units [4].

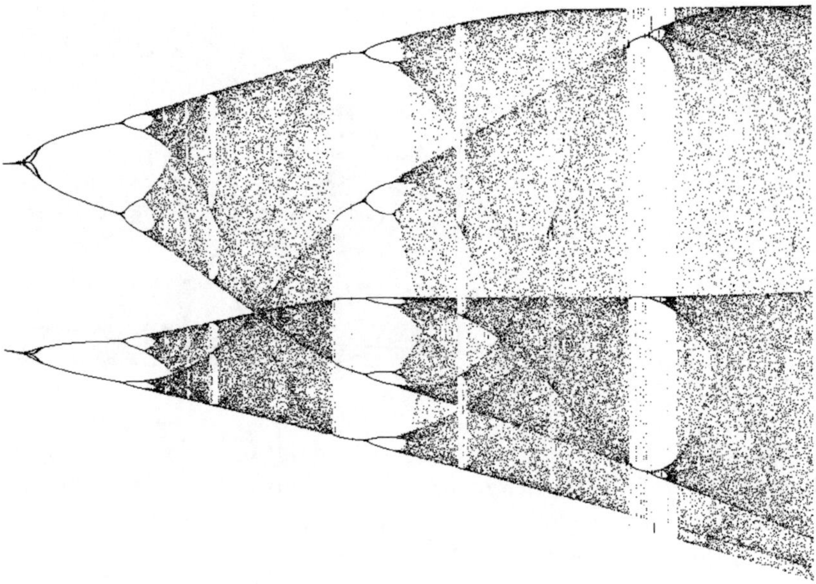

Figure 2. Intermittency and oscillatory behavior within a Feigenbaum Diagram.

Despite dramatic progress of modern reductionistic approaches in biological and medical research, there is a growing consensus that life processes need to be additionally interpreted from the above mentioned nonlinear point of view. One of these processes is cancer, a fundamental phenomenon of life and essential towards its deeper understanding.

Clocks and Chaos

Oscillations belong to the best known examples of self-organization in living organisms. Throughout the animal and plant kingdoms, oscillatory behavior is the rule rather than the exception [12]. Experimental evidence has finally emerged in confirmation of the idea that the frequency of a biological oscillator may encode dynamical information that controls basic cellular activity [30]. There is extensive evidence that clocks can contribute to create complexity and affect morphogenesis. Periodic reactions allow periodic signal transmitting between individual cells and regulate cell differentiation within an organism [17]. Such clocks are known to be extremely stable. The predictability of their behavior implies linearity. Under certain circumstances however, they may exhibit transition into chaotic behavior [15], implying unpredictability and nonlinearity.

Organisms have exploited the principles of self-organization by modifying and using, for example, oscillations as information-transducing signals, which can function as temporal control (clock) devices [26]. Such living systems as the well studied *dictyostelids* slime molds, as well as life as a whole, while existing far from thermodynamic equilibrium and displaying the properties of dissipative structures, do not commonly exhibit deterministic chaos behavior.

The cell cannot have a steady state unless it is accompanied by oscillations [2]. Chaos can easily be controlled by transition into an oscillatory state [27]. Thus, biological clocks are self-organization attractors and can naturally and experimentally prevent transition into deterministic chaos. Keeping chaos in check appears therefore to be an essential function of clocks.

The most commonly known biological clock is the circadian or “master clock” of vertebrates, located in the suprachiasmatic nucleus of the Hypothalamus (SCN), which allows synchronization to the periodic electromagnetic stimulus of sunlight. It is the most evident relationship between vertebrate temporal function and its environment. The master clock is driven by sunlight by means of excitation of the SCN. Periodic reactions build the basis of the master clock of organisms. They provide periodic signal transduction between single cells and regulate cell differentiation within an organism [16].

Clock impairment in general could consequently lead to impairment of intercellular periodic signal transduction and cellular differentiation, thus implying a self-organization break down leading to cell death or to cancer.

The Master Clock and Cancer

We previously observed that the activity of the master clock has periodic influence on systemic peroxidation as expressed by the serum levels of Malondialdehyde (MDA) of 39 healthy subjects. The phenomenon we found, which we termed “Periodic dip of Lipid peroxidation” (PDL), is apparently induced by the master clock’s activity because of its frequency and time of appearance at 03.00 h [6]. In a second trial [7], this phenomenon could not be detected in a group of 16 cancer patients in good general condition suffering from different solid malignant tumors in an advanced stage. The tests were performed at 3-h intervals over a 24-h period. The serum levels of Malondialdehyde (MDA) were examined by High-Performance Liquid Chromatography.

Figure 3 shows the mean normalized (transformed) 24-hour MDA levels in 3-hours intervals of the healthy subjects. A significant systemic phase of low lipid peroxidation with a frequency of once per 24 h was consistently present in 98% of the healthy subjects at 03.00 h.

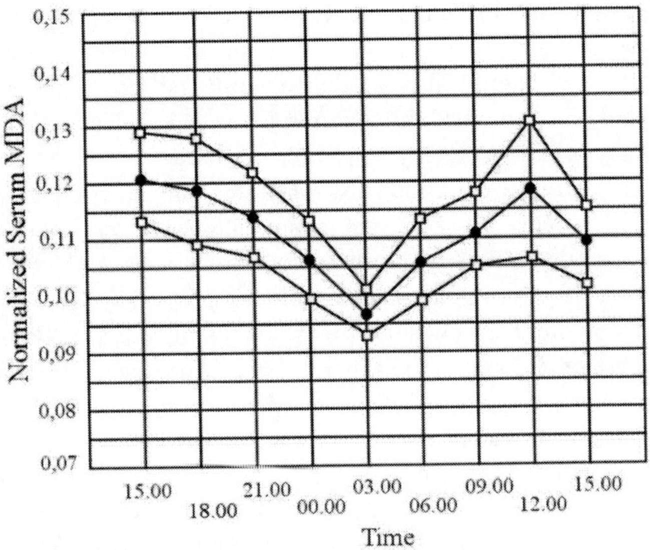

Figure 3. Mean normalized MDA levels of the healthy subjects (n=39) (dots). The squares show the 95% confidence interval.

Figure 4 shows the results of the cancer patients. The PDL could not be observed in any of the tested cancer patients. A significant difference between the 24-hour MDA levels of healthy subjects and cancer patients could only be found at 03.00 h ($p=0.000001$).

Bioenergetics, Evolution and Complexity: Back to the Beginning

Carcinogenesis generally features unique characteristics. A substantial part of potential cancer cells will not survive to ever be able to build a solid tumor, either due to selective mechanisms or to immune defenses. Their pathway towards malignancy is known to be extremely complex, hard and mostly prolonged. The required mutations to initiate it are considered to commonly appear by chance and if they ever attain further malignant development, cancer cells will be genetically extremely heterogeneous. Indeed, a solid malignant tumor could be regarded as a rather improbable incident.

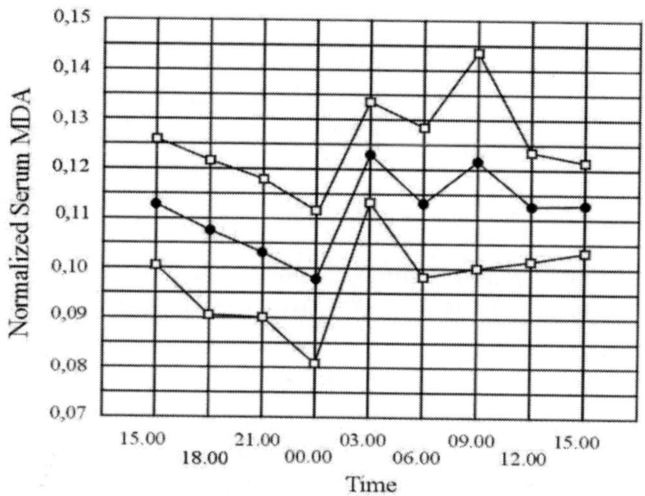

Figure 4. Mean normalized MDA levels of the test persons (n=39) (dots). The squares show the 95% confidence interval.

Due to their origin and genetic heterogeneity, cancer cells could additionally be expected to develop through many different pathways and render very many different results, which may not necessarily comprise malignancy. Yet, on one the hand cancer is an increasingly common disease. On the other hand, despite its apparent origin out of chance mutations and its heterogeneity, the outcome of carcinogenesis of solid tumors will almost always be quite similar: a growing and spreading tumor, its metastases and the microenvironment for which they are best adapted.

Furthermore, the development towards malignant tumors generally follows step by step a surprisingly stereotypic and deterministic strategy, e.g. intercellular signal disruption, gene suppression, gene amplification, proliferation, dedifferentiation, immortality, microenvironmental alteration, infiltration, immigration, angiogenesis, parasitism, destructiveness.

Along their pathway to malignancy, affected cells do not appear to be responding to selective pressure but rather to be remodeling their own microenvironment to fit their needs. It also appears to be that, every single time carcinogenesis is on its way, a specific entity within the organism is taking charge of orchestrating an extremely complex set of alterations out of almost any tissue to consistently deliver malignancy with its typical biologic behavior characteristics as its outcome.

As seen before, in open systems far from thermodynamic equilibrium, complexity spontaneously emerges giving rise to living nature. In our planet, sunlight stands as a fundamental energy source in the foremost place, paralleled by the existing environment to yield life. Multicellular organisms assimilate this energy directly from the sun or indirectly by nourishment. Metazoans have accomplished to optimize energy output by means of their mitochondria, the power plants of the cell, thus significantly increasing the distance of their systems from thermodynamic equilibrium, which enables them to reach a high degree of complexity.

It is widely believed that [mitochondria] arose from an endosymbiotic relationship between a glycolytic proto-eukaryotic cell and an oxidative bacterium [5] during the Precambrian eon. The onset of multicellular life, with its increasing degree of complexity as we know it today, probably took place on the basis of endosymbiosis. All present metazoan cells contain mitochondria.

Considering that mitochondria are the major reactive oxygen species (ROS) sources of the human body, we previously postulated [6], [7] that the observed phenomenon of a missing PDL in cancer patients could be related to mitochondrial respiratory activity, possibly down-regulated by NO· at 03.00 h. The presented results and the apparent failure of the master clock to function normally is obviously not the cause but rather a result of cancer. However, they could illustrate how clocks may be suppressed in the presence of an advanced malignant tumor. Should mitochondrial activity under physiologic conditions be indeed responsible for the function of the master clock as postulated above, our results could also illustrate, how mitochondrial function can be systemically impaired throughout a human body affected by a cancer disease. During the advanced phase of tumor growth cancer leads to profound alterations of host organs and functions. With the general reduction in energy available to the host, energy for seeking and processing food has almost vanished [10].

Mitochondrial ATP production normally accounts for ~80% of cellular energy [19]. However, 1-5% of the oxygen consumed by mitochondria is converted into reactive oxygen species (ROS). Mitochondria are therefore known to represent the major source of intracellular ROS, which can contribute to stimulate carcinogenesis. Unlike DNA, mtDNA lacks any histone protection against damage by ROS. Mitochondrial activity impairment itself is also known to increase ROS production, whereas excessive production of active oxygen species leads to mitochondrial dysfunction [17]. Mitochondria are further known to play a crucial part in apoptosis, one of the basic steps towards malignancy. Moreover, it is becoming clear that mitochondria are intimately involved in cellular Ca^{2+} signaling pathways [3], an essential function which could be impaired as a consequence of systemic mitochondrial degeneration.

Complex, multicellular life is known to have appeared in the form of the Ediacaran faunas about 550 million years ago. This event is termed the "Cambrian Explosion". Eukaryotes, single cells with differentiated nuclei and cell organelles, had already appeared 1.4 billion years before. There is indeed growing evidence for metazoan life occurring further back in the Precambrian than previously thought [21]. However, the environmental changes over the Cambrian/Precambrian boundary are potentially the driving forces behind the Cambrian Explosion event. Mitochondrial activity along with an adequate environment seem to represent the basis of the massive appearance of complex metazoan life. It is thought that a critical level of oxygen, along with adequate environmental conditions, may have been reached as a part of that environment.

Self-organizing systems are described as: dynamic, far from equilibrium, and absorbing energy to maintain their state [8]. Consequently, in a complex system crucially depending on its bioenergetics to maintain its distance from thermodynamic equilibrium, increasing impairment of mitochondrial activity of any origin should inevitably lead to a diminishing of that distance and eventually to a self-organization break down with a resulting loss of complexity.

Complexity, which has emerged and accumulated since Endosymbiosis, thus appears to progressively collapse in cancer cells. Impaired mitochondrial activity and altered microenvironmental changes thus appear to biologically transfer malignant cells to a state similar to the one existing previous to that event in the history of life on our planet. It has been therefore suggested that malignant cells develop back into their evolutionary past [13]. In the context of self-organization processes, this concept has previously been termed "Devolution" due to its antagonism to evolution [18]. The development of cancer cells towards malignancy could thus be observed as an adaptive survival response implementing a primitive pattern of life, which ATP delivered by glycolysis is still able to support. Hence, the orchestration of the stereotypic and deterministic development leading to cancer could be delivered by a set of preexisting archaic attractors, i.e. clock regulated genes stored in the memory of the system in a struggle for cell survival. Affected cells, due to their simplicity and modest energy requirements, may thus progressively turn into the fittest of the system the more the disease advances and the more its environment is altered. While the organism is increasingly affected by cachexia, the tumor and its metastases display sufficient robustness and energy to continue growing and infiltrating. Despite the irreversibility of nonlinear processes, cancer cells may accomplish evolutionary reversion or "devolution" by reactivating the oscillatory and genetic memory preserved in their systems to climb down the evolutionary stairs onto lower laminar phases and nonchaotic attractors, i.e. clocks once active eons ago to regulate gene expression in order to meet their survival needs.

Thus, a "devolutionary" concept of cancerogenesis would refer to a self-organization cell and microenvironmental remodelling where transformed cells progressively escape the collective properties of the organism by repressing their original set of physiologic attractors to evade their control function thus increasingly developing towards a predetermined set of archaic attractors and their corresponding genes. As a consequence, a cellular and systemic pattern with an inferior degree of complexity could be implemented, featuring the consistent characteristics of malignancy stored within the system.

It has been postulated, that natural selection is intimately linked to self-organization [4]. Environmental alterations contributing to malignancy may be due at least in part to peroxidation processes and chronic inflammation. A relationship between cancer and inflammation was proposed as early as 1847 by Rudolph Virchow [28]. H. Dvorak referred in 1986 to tumors as "wounds that do not heal" [11]. It has since become widely accepted that chronic inflammation provides a favorable environment for cancer to form and grow [24]. Tumor-infiltrating lymphocytes may contribute to cancer growth and spread, and to the immunosuppression associated with malignant disease [1].

Since Otto Warburg, Nobel award winner in 1931 [31], it is known that many tumors basically meet their energy requirements by means of aerobic glycolysis. Cancer-related mitochondrial defects include altered expression and activity of respiratory chain subunits and glycolytic enzymes, decreased oxidation of NADH-linked substrates, as well as mitochondrial DNA (mtDNA) mutations [5]. Alterations in mtDNA-encoded complexes impede oxidative phosphorylation and thus create variances in ATP synthesis, resulting in a cellular energy imbalance [9].

Extensive work has been done within the last decade in the field of chaos control. Part of it deals with the treatment of "dynamical diseases", to which cancer appears to belong, for example by means of excitation or suppression of oscillations, entrainment and synchronization, or transitions from chaotic to periodic oscillations and vice versa [29]. Simultaneous effect of red polarized light and liposomal solution of CoQ_{10} (LiQ-Sorb) supplementation re-energize mitochondrial ATP production via oxidative phosphorylation in brain, heart and skeletal muscle and affect coupling of oxidation and phosphorylation and regeneration of mitochondrial membrane integrity in 3-month-old rats in vivo [14]. Additionally, polarized light therapy had a beneficial effect on the healing of standardized wounds [23]. Finally, transcutaneous exposure of a small area of the human body to light with polychromatic visible and infrared (IR) polarized light leads to a fast decrease in the elevated pro-inflammatory cytokine plasma content and to an increase in the anti-inflammatory factor concentration, which may be an important mechanism of the anti-inflammatory effect of phototherapy [32]. These trials were performed on healthy humans or animals and may not deliver the same results in cancer patients. Additionally, the effectiveness of polarized light therapy in the context of cancer treatment would obviously depend on the uncertain reversibility of mitochondrial damage in cancer cells.

Polarized light may not be the solution for the cancer problem. But its mechanisms illustrate how physical stimuli could, for themselves or in addition to other treatment methods, contribute to regulate malignant cell behavior.

Conclusion

Malignant cells are unquestionably the fittest of an organism suffering from advanced malignancy. Excluding surgery, by trying to selectively kill those fittest cells of an affected organism, our present treatment methods are actually fighting against natural selection and may therefore not be successful in advanced cancer cases. Hence, it could be interesting to closely examine the possibility of allowing self-organization and selective pressure remodel cancer cells.

To achieve remodeling, original bioenergetics and microenvironmental conditions could be restored to enable self-organization to find its own way back to physiologic cell behavior. Cancer cells may not necessarily remain malignant if they are placed into a physiologic environment to which they could readapt and if they obtain the energy needed to recover their distance from thermodynamic equilibrium.

Our findings in the previously mentioned papers may illustrate how cancer cells could escape from their physiologic behavior by clock suppression. They may indicate as well, how reactivation of the original physiologic clocks and thus of the original self-organization patterns, i.e. cancer cell remodeling, could be the goal of a hypothetic cancer treatment.

The reductionistic approach to cancer has given us a wealth of knowledge about cancer diseases. Therapeutic advances achieved until the present are nonetheless insufficient because we still do not fundamentally know what the nature of cancer really is. Acquiring that knowledge by changing our concepts about cancer and its treatments could lead us to a causal and possibly more effective cancer therapy.

ACKNOWLEDGMENTS

I thank Dr. rer. nat. Wolfgang Bayer, the head of Laboratorium Dr. Bayer, Stuttgart, for implementing the MDA measurements. I also thank Dave Miller for his valuable comments and corrections. The Feigenbaum diagrams were created with the software program "Feigenbaum", generously provided by Ulrich Schwebinghaus. Figure 4 was reproduced by kind permission of Taylor & Francis Group <http://www.informaworld.com>. Original source of publication: Cardona, Mitochondria: The hands of the clock? *Biol Rhythm Res.,* 2006, 37, 21-28.

REFERENCES

[1] Balkwill, F; Mantovani, A. Inflammation and cancer: back to Virchow? *Lancet*, 2001, 357, 539-545.

[2] Bernhard, R. Survey of some biological aspects of irreversible dynamics. J Theor Biol, 1964, 3, 532-557.

[3] Briggs, J, Peat, FD, Turbulent Mirror, Harper & Row, New York, 1989.

[4] Burnett, P; Gaspers, SD; Thomas, AP. Control of Mitochondrial Metabolism by Calcium-Dependent Hormones. In: Lemasters JJ, Nieminen AL editors. *Mitochondria in Pathogenesis*. New York: Kluwer Akademic/Plenum; 2001, 153-175.

[5] Camazine, S; Deneubourg, J-L; Franks, NR; Sneyd, J; Theraulaz, G; Bonabeau, E. *Self-Organization in Biological Systems*. Princeton and Oxford: Princeton University Press; 2003.

[6] Carew, JS; Huang P. Mitochondrial defects in cancer. 2002. *Molecular Cancer*. URL: http://www.molecular-cancer.com/content/I/I/9

[7] Cardona, F. Mitochondria: The hands of the clock? *Biol Rhythm Res.,* 2006, 37, 21-28.

[8] Cardona, F. Periodic dip of lipid peroxidation in humans: a redox signal to synchronize peripheral circadian clocks? *Medical Hypotheses*, 2000, 463, 841-846.

[9] Cardona, F. Unpublished data.

[10] Coffey, D.S. Self-organization, complexity and chaos: The new biology for medicine. *Nat Med*, 1998, 8, 882-885.

[11] Copeland, WC; Wachsman, JT; Johnson, FM; Penta, JS. Mitochondrial DNA Alterations in Cancer. *Cancer Invest,* 2002, 20, 557-569.

[12] Costa, G. Cachexia, the metabolic component of neoplastic diseases. *Cancer Res*, 1977, 37, 2327-2335.

[13] Dvorak, HF. Tumors: wounds that do not heal. Similarities between tumor stroma generation and wound healing. *N Engl J Med*, 1986, 315, 1650-9.

[14] Edmunds, L. Series introduction. In: Rensing L, editor. *Oscillation and morphogenesis*. New York: Marcel Dekker; 1993, iii-iv.

[15] Greaves, M. *Krebs – der blinde Passagier der Evolution*. Heidelberg: Springer; 2000.

[16] Gvozdjáková, A; Kucharská, J; Pálinkás, J. Polarized Light and CoQ_{10} Effect in Mitochondria (Pre-Clinical Study). In: Gvozdáková A editor. *Mitochondrial Medicine, Mitochondrial Metabolism, Diseases, Diagnosis and Therapy*. New York: Springer Science + Bussiness Media; 2008, 396-398.

[17] Hess, B; Markus, M. Chemische Uhren. In: Dress A, Hendrichs H, Küppers G editors. *Selbstorganisation, Die Entstehung von Ordnung in Natur und Gesellschaft*. München: Piper; 1986, 61-79.

[18] Hess, B; Markus, M. Ordnung und Chaos in chemischen Uhren. In: Küpper BO editor. *Ordnung aus dem Chaos, Prinzipien der Selbstorganisation und Evolution des Lebens*. München: Piper; 1987, 157-174.

[19] Higuchi, H; Hiromasa, I. Mitochondrial Changes after Acute Alcohol Ingestion. In: Lemasters JJ, Nieminen AL editors. *Mitochondria in Pathogenesis*. New York: Kluwer Academic/Plenum; 2001, 379-391.

[20] Karu, T, Pyatibrat, L, Kalendo, G, Irradiation with He-Ne laser increases ATP level in cells cultivated in vitro. *J Photochem Photobiol B*. 1995, 27, 219-23.

[21] Karu TI, Mitochondrial signaling in mammalian cells activated by red and near-IR radiation. *Photochem Photobiol*, 2008, 84,1091-9.

[22] Kang, D; Hamasaki, N. Mitochondrial Oxidative Stress and Mitochondrial DNA. Clin Chem Lab Med, 2003, 41, 1281-1288.

[23] Kraus, M; Wolf, B. Emergence of Self-Organization in Tumor Cells: Relevance for Diagnosis and Therapy. *Tumor Biology,* 1993, 14, 338-353.

[24] Lane, A. The Cambrian Explosion. 1999. URL: http://palaeo.gly.bris.ac.uk/Palaeofiles/Cambrian/timing/palaeo/palaeo.html

[25] Liening, A., *Komplexe Systeme zwischen Ordnung und Chaos: Neuere Entwicklungen in der Theorie nicht-linearer dynamischer Systeme und die*

Bedeutung für die Wirtschaftswissenschaft und ihre Didaktik. München: LIT Verlag, 1999.

[26] Monstrey, S; Hoeksema, H; Depuydt, K; Van Maele, G; Van Landuyt, K; Blondeel, P. The effect of polarized light on wound healing. *European Journal of Plastic Surgery,* 2004, 24, 377-382.

[27] Nam, JH; Murphy, S. Chronic Inflammation and cancer in various organ systems. In: Morgan DW, Forssmann UJ, Nakada MT editors. *Cancer and Inflammation*. Basel: Birkhäuser; 2004, 1-20.

[28] Peitgen, HO, Jürgens, H, Saupe D. Chaos – Bausteine der Natur, Springer-Verlag, 1994.

[29] Prigogine, I; Stengers I. *Dialog mit der Natur*. 5. Auflage. München: Piper; 1986.

[30] Rensing, L; Deutsch, A. Introduction and Overview. In: Rensing L editor. *Oscillations and Morphogenesis*. New York: Marcel Dekker; 1993, 1-23.

[31] Setälä, K. Carcinogenesis – Devolution Towards An Ancient Nucleated Pre-Eukaryotic Level, *Medical Hypotheses*, 1984, 15, 209-230.

[32] Schneider, FW; Münster, AF. *Nicht Lineare Dynamik in der Chemie*. Heidelberg: Spektrum Akademischer Verlag; 1996.

[33] Schulz, TJ, Thierbach, R, Voigt, A, Drewes, G, Mietzner, B, Steinberg, P, Pfeiffer, AF, Ristow, M, Induction of oxidative metabolism by mitochondrial frataxin inhibits cancer growth: Otto Warburg revisited. J Biol Chem. 2006, 281, 977-81.

[34] Taylor, BL, Zhulin, IB, PAS Domains: Internal Sensors of Oxygen, Redox Potential, and Light, Microbiol Mol Biol Rev, 1999, 63, 479–506.

[35] Virchow, R. Zur Entwicklung des Krebses. *Arch path Anat Physiol*, 1847, 1, 94-201.

[36] Walleczek, J. The Frontiers and challenges of biodynamics research. In: Walleczek J editor. *Self-Organized Biological Dynamics & Nonlinear Control.* Cambridge: Cambridge University Press; 2000, 1-11.

[37] Walleczek, J; Eichwald, CF. Enzyme kinetics and nonlinear biochemical amplification in response to static and oscillating magnetic fields. In: Walleczek, J editor. *Self-Organized Biological Dynamics & Nonlinear Control*. Cambridge: Cambridge University Press; 2000, 193-215.

[38] Warburg, O. On the Origin of Cancer Cells, *Science*. 1956, 123, 309-314.

[39] Zhevago, NA; Samoilova, KA. Pro- and anti-inflammatory cytokine content in human peripheral blood after its transcutaneous (in vivo) and direct (in vitro) irradiation with polychromatic visible and infrared light. *Photomed Laser Surg,* 2006, 24, 129-39.

In: Biological Clocks: Effects on Behavior… ISBN: 978-1-60741-251-9
Editors: O. Salvenmoser et al. pp. 177-190

Chapter 8

KLEPSYDRAIC MODEL OF INTERNAL TIME REPRESENTATION: EXPERIMENTAL FINDINGS AND ANALYTICAL PROPERTIES

Jiří Wackermann
Dept. of Empirical and Analytical Psychophysics, Institute for Frontier Areas of Psychology and Mental Health, Freiburg i.Br., Germany

ABSTRACT

In this chapter a "klepsydraic" model of internal time representation is introduced. The model's working is illustrated by experimental data on duration reproduction and duration discrimination. In addition, special properties of the reproduction function predicted by the model are discussed. It is shown that time-scales generated by "klepsydraic clocks," although non-uniform with respect to the objective (physical) time, allow for rational time-keeping. These findings recommend the model as an analytic instrument in studies of time perception, timing behavior and subjective time awareness.

INTRODUCTION

Of our interest is perception of temporal intervals in the supra-second range, that is, in the order of magnitudes 10^0—10^1 sec. This domain is relatively under-represented in the literature. Most of the current experimental research into "temporal cognition" is being done in the sub-second range, whereas chronobiology traditionally focusses on ultradian and circadian rhythms. However, time perception and timing behavior in the supra-second domain are of importance for purposeful action in many contexts, *e.g.* locomotor activities in environmental space (sports, traffic, etc.). Also importantly, subjective experience of time as a perceived extension is fully developed only in the range of our interest; shorter or longer durations are rather cognitive constructs. Studies in supra-second domain thus may also contribute to our understanding of time awareness as a fundamental dimension of human consciousness.

Internal Time Representation

Cognition of temporal relations between environmental events, and purposeful behavior informed by these relations, requires a kind of intra-organismic (presumably neural) representation of temporal data. Still, our knowledge of the underlying neurobiological principles and mechanisms is scarce, in spite of increasing amount of particular findings [1, 2].

Contemporary theories and models of internal time representations are mostly based on a pacemaker–counter principle, proposed almost half-century ago [3]. In this scheme, discrete signals ("pulses") are emitted by a central instance ("pacemaker") and gated to an accumulator ("counter"); the pulse count accumulated during a time interval serves as an internal measure of its duration. This is but one among many possible mechanisms of internal time-keeping, which all could be by equal right called "internal clocks." However, during a few past decades this scheme has been adopted as *the* "internal clock model" (ICM) and widely accepted as a principle of internal time representation in psychological science [4, 5] as well as in neurobiology [6].

The dominant rôle of the ICM manifests itself in the fact that empirical findings in time perception and timing behavior are often presented in the ICM - based language of "pacemaker frequency" and "pulses", even if there is no necessity to postulate such entities or processes, and no observational evidence for them. In this way, the model has been degraded to merely a *façon de parler*, insinuating that time-related cognition and action require a sort of ticking mechanism [7].

In our approach, durations are represented by continuous state variables, so we do not assume discrete "pulses" or "ticks." The states can be compared only in terms of primitive ordinal relations, 'more', 'less', or 'equal'; the internal representation is, in this sense, of "pre-metric" nature. Hence the elementary operation of temporal cognition is comparison of time intervals, not a conversion of a single interval to a numerical datum or *vice versa*.

DUAL KLEPSYDRA MODEL

The "klepsydraic" model [8, 9] is based on an internal representation of time extension by a state of an inflow–outflow unit (IOU) resulting from integration (continuous summation) of an internal "flow" in a "leaky accumulator."[1] For perception of two subsequently presented time intervals in a reproduction or pairwise comparison experiment, two IOUs are allocated and their states continuously compared (dual klepsydra model, DKM). Equality of states of the two accumulators is subjectively experienced as equality of the two durations.

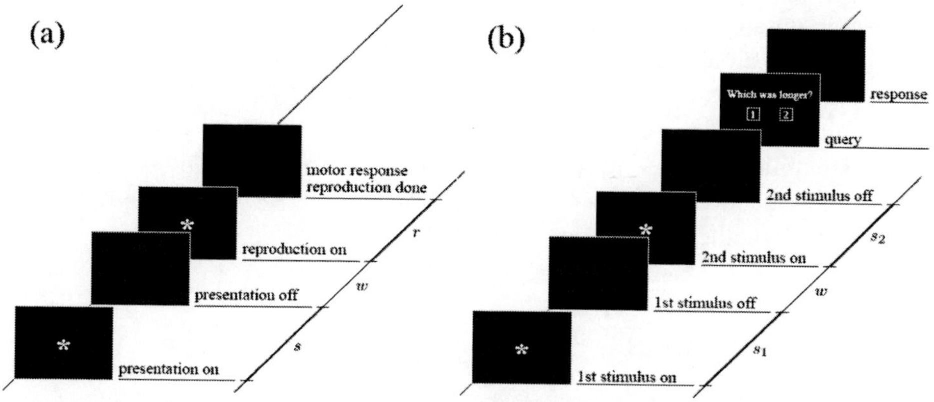

Figure 1. Two principal paradigms used in duration discrimination experiments. Time intervals are marked by sensorily perceivable (*e.g.* visual) stimuli serving as "duration carriers." (a) Reproduction task: s and w are determined by the experimental protocol, r by the subject's response. (b) Pairwise comparison task: s_1, s_2 and w are determined by the experimental protocol.

Let us assume linear IOUs, governed by a differential equation

$$\mathrm{d}Y = (i - \kappa\, Y)\, \mathrm{d}t, \tag{1}$$

[1] Hence the name of the model: Greek *klepsydra* = water clock.

where Y is the accumulated state, i is the unit's inflow, and κ is a coefficient relating the outflow rate to the momentary state. Then the DKM predicts the subjectively equal *reproduced* duration, r, to be

$$r = \mathrm{krf}(s,w) = \kappa^{-1} \ln(1 + \eta\,(1 - e^{-\kappa s})\, e^{-\kappa w}), \tag{2}$$

where s is the presented duration and w is the inter-stimulus interval between the end of the presentation phase and the beginning of the reproduction phase (*cf.* Figure 1a). The form of the "klepsydraic reproduction function" (KRF: eq. 2) depends on two parameters, the loss rate parameter κ, which is assumed to be identical for both units, and the inflow ratio $\eta = i_1/i_2$ (see eq. 1). For $\kappa \rightarrow 0$, the KRF becomes a linear function of s independent of w, $\mathrm{krf}(s,w) = \eta\, s$.

In a deterministic version of the model (DKM), the flows are piecewise constant functions of time. In a stochastic version of the model (SDKM) [10, 11], the flows are subject to random fluctuations, modeled by superimposed white noise with variance σ^2; this model accounts for intra-individual dispersion of responses in duration reproduction/discrimination experiments.

The DKM is an abstract, mathematical model; in a neurobiologically realistic model, the processes of accumulation *vs.* spontaneous loss of accumulated state may be interpreted as excitation *vs.* de-excitation (or active inhibition) of local neural assemblies [12]. A functional MRI study of brain activation during the duration reproduction task (see below) revealed a bi-focal activation pattern similar to the hypothetical DKM mechanism [13].

Experimental Findings

Here we refer to data on discrimination of temporal durations, obtained in a reproduction or pairwise comparison paradigm.[2] In the reproduction task (Figure 1a), the subject indicates by a motor action (button press) the moment when the second interval (duration r) equals, in her/his perception, to the first interval (duration s). In the pairwise comparison task, the subject perceives two successive intervals of durations s_1 and s_2, and gives a judgment of which interval was longer or shorter. A "point of subjective equality" (PSE) $({}^{\circ}s_1, {}^{\circ}s_2)$ is then determined from

[2] The "interval production" or "magnitude estimation" paradigms, popular in time perception research, are clearly not suitable for our purpose, as they require conversion of perceived durations to conventional units, and thus involve the subject's memory for units and understanding of arithmetics.

the statistics of responses (*cf.* Figure 3). According to the DKM, the equality $°s_2 = \mathrm{krf}(°s_1,w)$ should hold.

Data from duration reproduction experiments (individual averages over a sufficient number of repeated trials, or group averages, *i.e.*, "grand means") usually result in a monotonically increasing response function with a negative curvature (Figure 2). This form corresponds to the phenomenon of "subjective shortening in memory" of past temporal intervals [14], which is a *non-linear* function of s: the reproduced-to-presented duration ratio, r/s, is itself a decreasing function of s. The reproduction response curves are fitted by the KRF with a very good accuracy. Estimates of the parameter κ are usually in the range $1—4 \times 10^{-2}$ sec [9, 15]; the corresponding relaxation times of the leaky accumulators are proportional to κ^{-1}, and thus in the range approx. 20—60 seconds.

Data from discrimination of pairwise presented durations typically show a shift of the PSE toward the ratio $°s_2/°s_1 < 1$ (Figure 3). This asymmetry of the discrimination function [16] reflects the lossy character of the internal representation of past durations, similarly to the progressive shortening of the reproduction response. Estimates of κ derived from individual PSEs are in the same order of magnitudes as those from reproduction data, indicating that both phenomena belong to the same mechanism. In addition to κ, we can estimate the SDKM parameter $\gamma \equiv i/\sigma$ from the slope of the psychometric function at the PSE. Estimates of the "diffusion time" γ^{-2} from our duration discrimination data are in the range 150—300 msec [17].

We have hypothesized that the parameter κ may reflect the global organismic state [9]. Two groups of experimental results, based on intra-individually repeated duration reproduction experiments, support this conjecture. In series of repeated reproduction tasks during the day, κ co-varied with the individual phase of the circadian cycle, assessed with respect to (w.r.t.) to the individual's sleep period [15]. In series of reproduction tasks after administration of a psychotropic drug, psilocybin, κ was significantly increased at times of the maximum drug effect [18]. The latter results may shed light on drug-induced changes of subjective experience of time.

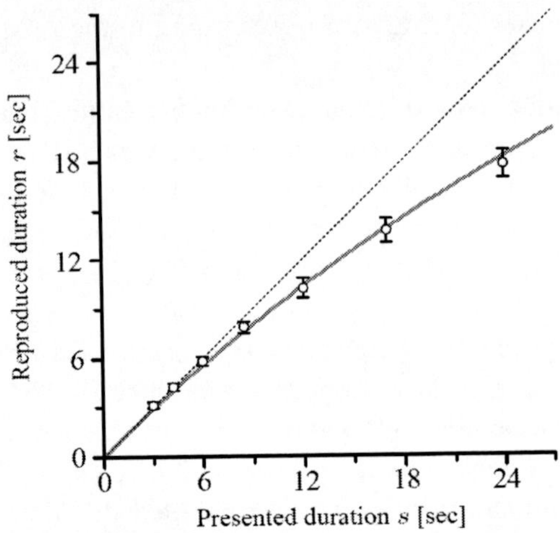

Figure 2. Average responses from a duration reproduction experiment, aggregated over 12 subjects and 20—30 trials/subject. Circles: grand means; error bars: ±1 st. dev. of the grand mean; dotted line: chronometrically correct response $r = s$; grey curve: fitted KRF with $\kappa = 0.013\ \mathrm{sec}^{-1}$ and $\eta = 1$. — Reprinted from [20], with permission from Imprint Academic.

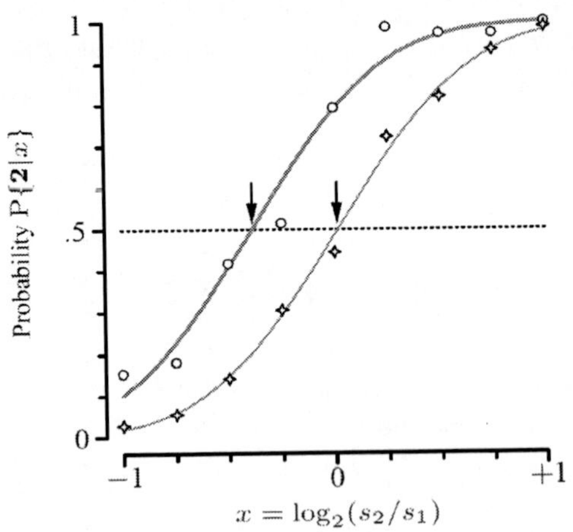

Figure 3. Average responses from a pairwise comparison experiment, aggregated over 9 subjects and 8 trials/subject for each data point. Circles/rhomboids: grand means of relative frequency of response "2nd interval longer". Duration pairs (s_1,s_2) were varied on nine levels at a constant geometric mean $\sqrt{(s_1\,s_2)} = m$, with $m = 3$ sec (rhomboids) and $m = 6$ sec (circles). Grey curves: Gaussian "psychometric functions" fitted separately to data sets for $m = 3$ and 6 sec. Intersections with $P = ½$ (dotted line) determine the PSEs (arrows). — Based on data from [16], adapted from [20], with permission from Imprint Academic.

Analytical Properties

The klepsydraic reproduction function (eq. 2) is not merely an apt data-fitting formula; it also has interesting properties worth further examination.

We start from an observation of "reproductive additivity." Consider a time interval composed of two adjacent subintervals s_1 and s_2 (Figure 4). Subsequent klepsydraic reproduction of the two parts,

$$r_1 = \text{krf}(s_1, s_2 + w),\ r_2 = \text{krf}(s_2, w + r_1),$$

results in a total identical to the reproduction of the whole, undivided interval

$$r_1 + r_2 = \text{krf}(s_1 + s_2, w).$$

This is similar to additivity property normally known (and required) in measurements of physical extensions; here, however, the measure (*i.e.*, the reproduction response) is dependent also on the interval separating the perceived and the reproduced interval.

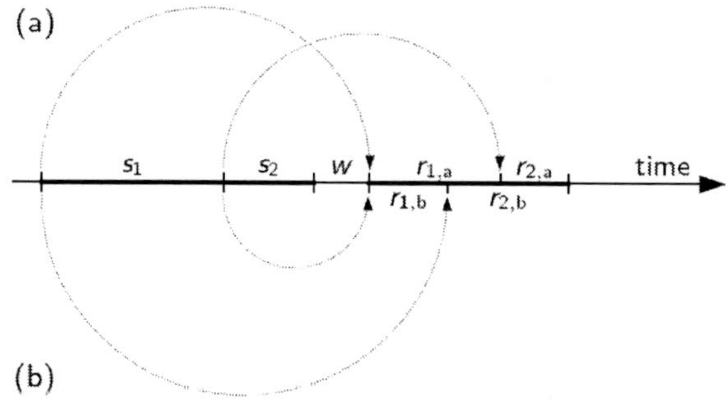

Figure 4. Additivity of klepsydraic reproduction of time intervals. Reproduction by parts, $s_1 \rightarrow r_1$, $s_2 \rightarrow r_2$, yields the sum $r_1 + r_2$ equal to reproduction of the entire duration $s_1 + s_2$ (a). The same is true for reproduction in reverse order, $s_2 \rightarrow r_1$, $s_1 \rightarrow r_2$ (b), although the reproduced times are generally not equal, $r_{1,a} \neq r_{2,b}$, $r_{2,a} \neq r_{1,b}$. — Reprinted from [19], with permission from Elsevier Science.

To formalize this notion, we define "serial additivity" [19] consistently with the above-stated property, generalized for n-partitions (with any $n > 1$) of temporal intervals, reproduced *per partes* in an arbitrary order. The general solution of functional equations specifying the "serial additivity" is the KRF, involving, as special cases, a linear response ($\kappa = 0$) and a constant null response

($\eta = 0$). This is a strong result suggesting the KRF as an "ideal form of the law of duration reproduction" [19].

Klepsydraic Clock and Klepsydraic Time-Scales

Measurement of a spatial or temporal extension consists in an apposition a series of equal units to the measured interval; the number of units needed to cover the interval provides its measure. In the case of time measurement, the apposited series is generated by a (by convention "uniform") reproduction mechanism, *e.g.* a swinging pendulum. However, any deterministic interval-reproduction mechanism generates a time-scale that can be used to measure other durations.

Of interest are now properties of time-scales generated by a klepsydraic clock (KC), based on iterated reproduction of a unit duration u by the KRF:

$$t_n = \Sigma_{k=1\ldots n}\, r_k \,,$$

where $r_k = \mathrm{krf}(r_{k-1},0)$, $r_0 \equiv u$. An important property of the KC is *reproductive consistency*: $\mathrm{krf}(r_n, 0) = \mathrm{krf}(u, t_n)$ for any n.

Due to the lossy klepsydraic representation, the subsequent KC-periods r_k become progressively shorter, so that the KC-generated series is non-uniform w.r.t. a regular (*e.g.* pendulum driven) clock. Or, inversely, the *horological function* of thc KC,

$$\vartheta(t) = \varepsilon^{-1}\,(\mathrm{e}^{\kappa t} - 1)\,,$$

where $\varepsilon \equiv 1 - \mathrm{e}^{-\kappa u}$, is a non-linear function of t. For a KC started at time t^{o}, the measure of an interval [t', t"] is

$$\vartheta(t'' - t^{\mathrm{o}}) - \vartheta(t' - t^{\mathrm{o}}) \qquad (3)$$

and thus depends on time instant t^{o} ("clock's epoch"). This, however, does not preclude the possibility of a rational time-keeping by klepsydraic clocks. Expression (3) can be factorized as

$$\exp(-\kappa t^{\mathrm{o}})\, \exp \kappa t' \;\vartheta(t'' - t')\,,$$

and thus the *chronometric ratio* between two KCs, i and j,

$$\lambda_{ij} \equiv (\vartheta(t'' - t^o_i) - \vartheta(t' - t^o_i)) / (\vartheta(t'' - t^o_j) - \vartheta(t' - t^o_j)) = \exp \kappa (t^o_j - t^o_i)$$

is invariant under the interval's translation or dilatation.[3] It follows that readings $T_i \equiv \vartheta(t - t^o_i)$ of KCs $i = 1, 2, 3...$, started at different epochs t^o_i, are at any time t related by affine transformations

$$T_i = \mathsf{A}_{ij}(T_j) = \lambda_{ij} T_j + \mu_{ij} \tag{4}$$

(Figure 5). Transformations A_{ij} naturally form a group: $\mathsf{A}_{ii} = \text{id}$, $\mathsf{A}_{ij} = \mathsf{A}_{ij}^{-1}$, $\mathsf{A}_{ij}\,\mathsf{A}_{jk} = \mathsf{A}_{ik}$. Therefore, time-scales generated by an ensemble of KCs all operating at the same loss-rate κ, but (generally) with different units u_i and started at different epochs t^o_i, are mutually "uniform", though they are non-uniform w.r.t. the pendulum clock time-scale. Such time-scales may be called "quasi-uniform" [20]. As seen above, quasi-uniformity is sufficient for a rational time-keeping: an ensemble of cognitive agents (subjects, observers) equipped with KCs can agree upon an inter-subjectively consistent description of the world (that is, a kind of "physics") in terms of temporal relations.[4]

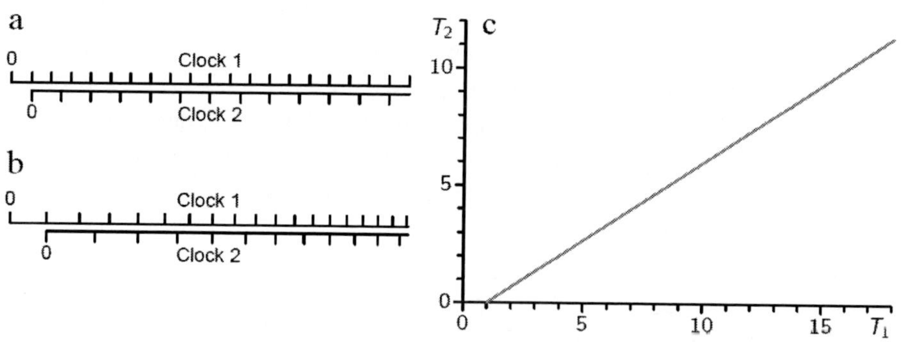

Figure 5. Transformations between clock readings. (a) Two uniform tick series. (b) Two klepsydraic tick series (κ = 0.1). In both cases, (a) and (b), the clock readings are transformed by eq. (4) with λ_{12} = 3/2, μ_{12} = 1. The chronometric ratio λ_{12} is constant although the periods of klepsydraic clocks are not.

---

[3] This can be easily generalized for the case of clocks operating with different units, u_i and u_j ; the chronometric ratio λ_{ij} then involves the quotient $\varepsilon_i/\varepsilon_j$.

[4] There is an interesting parallel between this property of KC-generated time-scales and the intepretation of red-shift of light from distant sources in Milne's cosmological model [21]: in both cases, we observe a consequence of "time unit aging" in clocks started in the past; *cf.* Appendix to [20].

Subjective Awareness of Duration and Temporal Order

Temporal order of events, and of temporal extensions separating the events, are constitutive moments of human consciousness.[5] The subjective awareness of time, in a narrower sense, is bound to "duration" as a perceptual datum. Interestingly, the range of perceivable temporal extensions is restricted to an interval from fractions of a second to a few tens of seconds. Durations outside of these limits, or "horizons" [22], are cognitively accessible, that is, *known*, but *not really perceived*. In our model, the "outer" horizon is determined by the relaxation time (κ^{-1}) of the leaky accumulators, whereas the "inner" horizon is determined by the diffusion time (γ^{-2}), depending on the signal-to-noise ratio of the inner flows.

Analysis of conditions for a retrospective reconstruction of temporal order from klepsydraically represented durations [23] reveals an interplay between the two parameters, where the dimensionless ratio of the relaxation and diffusion time, $C \equiv \gamma^2/\kappa$, plays the rôle of critical parameter. As long as $C \gg 1$, the two horizons are kept apart, leaving a range of sufficient discrimination acuity for the reconstruction of temporal succession (Figure 6). With κ increasing and/or γ decreasing, as the ratio C goes to ≈ 1, the two horizons approach each other, and may eventually fusionate, so that a reliable cognition of the succession of past events is impossible. This phenomenon may be related to alterations of time awareness reported from so-called altered states of consciousness, such as "time standing still": if temporal order is abolished, a cognitive consequence is that all events present to the individual's consciousness are given in a "permanent presence," "eternal now."

Directions of Further Research

There is a cumulative evidence for a modulating rôle being played by different neurotransmitter systems, serotoninergic and dopaminergic, in the internal time representation [24, 1]. Effects of a serotoninergic hallucinogen, psilocybin, on duration reproduction [25] could be interpreted within the DKM framework as changes of the klepsydraic loss rate parameter, κ [18]. More experimental studies are needed to explore reactivity of κ to diverse neurochemical agents. An important question is how far these findings can be transfered from the level of inter-individual, state-dependent effects to stable

[5] A fact that gave rise to a questionable notion of "psychological time"; *cf.* Section 6.1 in [20].

features, manifested in inter-individual or inter-group differences in time perception, possibly on a genetic basis [26]. A multidimensional cooperative study (work in progress) focussing on differences in duration reproduction and discrimination, and their correlations with brain electrical activity, in sub-populations characterized in terms of polymorphism of relevant genes should elucidate this question.

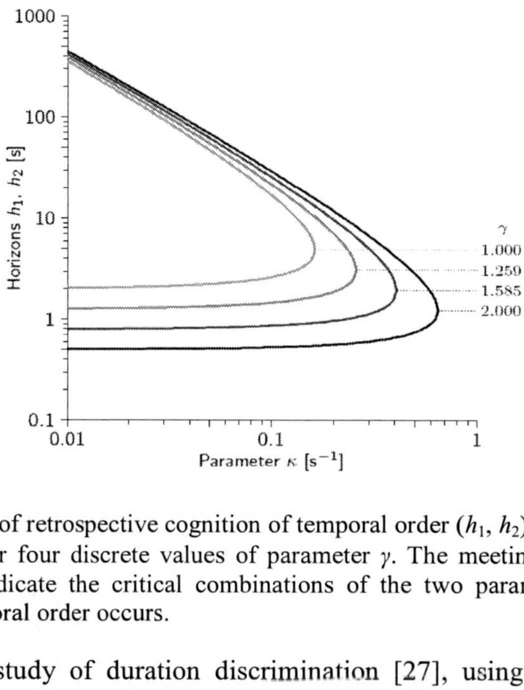

Figure 6. Horizons of retrospective cognition of temporal order (h_1, h_2) plotted as functions of parameter κ for four discrete values of parameter γ. The meeting points of the two branches h_1, h_2 indicate the critical combinations of the two parameters, for which a "collapse" of temporal order occurs.

In a recent study of duration discrimination [27], using visual stimuli of varied luminosity, a paradoxical interaction between the stimulus brightness and subjective duration was found: time intervals marked by *brighter* stimuli were on the average perceived as *shorter*. This inter-modal effect was superimposed on the unspecific "subjective shortening" effect, manifested by the PSE shift and explained by the lossy character of the internal representation [16]. Pairs of dim–bright and bright–dim stimuli were intermixed within one experimental session, so that we may assume identical loss rate (κ) for all trials, and attribute the cross-modal effect to a modulation of internal flows (η). Thus it occurs possible to separate experimentally effects assessed by the two DKM parameters, κ and η. Future work will aim at securing the observed effect in larger data sets, and at generalizing the effect for other physical properties of stimuli used as “duration carriers.”

CONCLUSION

The dual klepsydra model (DKM) is a useful analytic instrument in studies of human time perception in the supra-second region. The “klepsydraic reproduction function” (KRF) predicted by the model fits very well data from duration reproduction experiments. In particular, the model accounts for the following phenomena:

- progressive shortening of reproduction response with increasing presented durations;
- asymmetry of the discrimination function in pairwise comparison of durations;
- the “inner” and “outer” horizons delimiting perceived duration and temporal order.

In addition, the KRF has a special property, serial additivity, making it an “ideal law of duration reproduction.” Time-scales generated by “klepsydraic clocks” are not uniform w.r.t. physical time, but they are “quasi-uniform” in terms of mutual transformations, and thus allow for a rational timekeeping.

The DKM has been used successfully to parametrize data in studies of organismic or environmental factors affecting time perception and/or subjective time awareness. All these facts recommend the klepsydraic model or, more generally, the klepsydraic principle of duration representation to the attention of researchers studying time perception, timing behavior, or subjective time awareness.

ACKNOWLEDGMENTS

Thanks to Carsten Allefeld for helpful comments on an earlier draft of this chapter.

REFERENCES

Mauk, M.D. & Buonomano, D.V. (2004). The neural basis of temporal processing. *Annual Review of Neuroscience*, *27*, 307–340.

Rubia, K. & Smith, A. (2004). The neural correlates of cognitive time management: a review. *Acta Neurobiologiae Experimentalis*, *64*, 329–340.

Treisman, M. (1963). Temporal discrimination and the indifference interval. Implications for a model of the 'internal clock'. *Psychological Monographs*, *77*, 1–31.

Zakay, D. & Block, R.A. (1997). Temporal cognition. *Current Directions in Psychological Science*, *6*, 12–16.

Grondin, S. (2001). From physical time to the first and second moments of psychological time. *Psychological Bulletin*, *127*, 22–44.

Meck, W. H. (2005). Neuropsychology of timing and time perception. *Brain and Cognition*, *58*, 1–8.

Buhusi, C. V. & Meck, W. H. (2005). What makes us tick? Functional and neural mechanisms of interval timing. *Nature Reviews of Neuroscience*, *6*, 755–765.

Wackermann, J., Ehm, W. & Späti, J. (2003). The 'klepsydra' model of internal time representation. In *Fechner Day 2003* (ed. by B. Berglund & E. Borg), Stockholm: International Society for Psychophysics, pp. 331–336.

Wackermann, J. & Ehm, W. (2006). The dual klepsydra model of internal time representation and time reproduction. *Journal of Theoretical Biology*, *239*, 482–493.

Ehm, W. & Wackermann, J. (2003). Analysis of a stochastic version of the 'klepsydra model'. In *Fechner Day 2003* (ed. by B. Berglund & E. Borg), Stockholm: International Society for Psychophysics, pp. 59–64.

Ehm, W. & Wackermann, J. (2004). Estimating the parameters of the stochastic klepsydra model. In *Fechner Day 2004* (ed. by A. Oliveira, M. Teixeira, G. Borges, & M. Ferro), Coimbra: International Society for Psychophysics, pp. 350–355.

Wackermann, J. (2005). Experience of time passage: Phenomenology, psychophysics, and biophysical modelling. In *Endophysics, Time, Quantum, and the Subjective* (ed. by R. Buccheri, M. Saniga, & A. Elitzur), Singapore: World Scientific, pp. 189–208.

Jech, R., Dušek, P., Wackermann, J., & Vymazal, J. (2005). Cumulative blood oxygenation level-dependent signal changes support the 'time accumulator' hypothesis. *Neuroreport*, *16*, 1467–1471.

Wearden, J.H. & Ferrara, A. (1993). Subjective shortening in human's memory for stimulus duration. *Quarterly Journal of Experimental Psychology, 46B*, 1–25.

Späti, J.P. (2005). *Time perception in the context of circadian activity.* Diploma thesis, Faculty of Biology, Albert-Ludwigs-University, Freiburg.

Wackermann, J. & Späti, J. (2006). Asymmetry of the discrimination function for temporal durations in human subjects. *Acta Neurobiologiae Experimentalis*, *66*, 245–254.

Wackermann, J. & Ehm, W. (2007). Dual klepsydra model of duration discrimination. In: *Fechner Day, 2007* (ed. by S. Mori, T. Miyaoka, & W. Wong), Tokyo: International Society for Psychophysics, pp. 515–520.

Wackermann, J., Wittmann, M., Hasler, F., & Vollenweider, F.X. (2008). Effects of varied doses of psilocybin on time interval reproduction in human subjects. *Neuroscience Letters*, *435*, 51–55.

Wackermann, J. (2006). On additivity of duration reproduction functions. *Journal of Mathematical Psychology*, *50*, 495–500.

Wackermann, J. (2008). Measure of time: a meeting point of psychophysics and fundamental physics. *Mind and Matter*, *6(1)*, 9–50.

Whitrow, G.J. (1996). E.A. Milne and cosmology. *Quarterly Journal of the Royal Astronomical Society*, *37*, 365–367.

Wackermann, J. (2007). Inner and outer horizons of time experience. *Spanish Journal of Psychology*, *10*, 20–32.

Wackermann, J. (2007). A psychophysical model for retrospective temporal order. Paper read at the 14th Autumn Academy "*Theory in Cognitive Neuroscience*," Wildbad Kreuth, November 2007.

Rammsayer, T. H. (1999). Neuropharmacological evidence for different timing mechanisms in humans. *Quarterly Journal of Experimental Psychology*, *52B*, 273–286.

Wittmann, M., Carter, O., Hasler, F., Cahn, B.R., Grimberg, U., Spring, P., Hell, D., Flohr, H., & Vollenweider, F.X. (2007). Effects of psilocybin on time perception and temporal control of behaviour in humans. *Journal of Psychopharmacology*, *21*, 50–64.

Sysoeva, O.V., Portnova, G.V., Maluchenko, N.V., Tonevitsky, A.G., & Ivanitsky, A.M. (2008). Inter-individual differences in time perception: Gene-brain-psychological function. *International Journal of Psychophysiology*, *69*, 189–190.

Wackermann, J. & Meyer-Blankenburg, N. (2009). Differential effect of stimulus brightness in visual duration discrimination. In *Fechner Day 2009* (ed. by M.A. Elliott, S. Antonijević, S. Berthaud, P. Mulcahy, B. Bargary, C.

Martyn, & H. Schmidt), Galway: International Society for Psychophysics, pp. 459–464.

INDEX

A

B

C

D

E

F

G

H

I

J

K

L

M

N

O

P

Q

R

S

T